# SCIENTIFIC AMERICAN

### PRESENTS

# NOBEL PRIZE WINNERS

## ON

# MEDICINE

PUBLISHING

New York

© 2009 Scientific American

Published by Kaplan Publishing, a division of Kaplan, Inc.
1 Liberty Plaza, 24th Floor
New York, NY 10006

Library of Congress Cataloging-in-Publication Data has been applied for.

Printed in the United States of America

10 9 8 7 6 5 4 3 2 1

ISBN: 978-1-60714-468-7

Kaplan Publishing books are available at special quantity discounts to use for sales promotions, employee premiums, or educational purposes. Please email our Special Sales Department to order or for more information at kaplanpublishing@kaplan.com, or write to Kaplan Publishing, 1 Liberty Plaza, 24th Floor, New York, NY 10006.

# Table of Contents

# TABLE OF CONTENTS

# TABLE OF CONTENTS

The number in parentheses is the year
in which the author won the Nobel Prize.

# Muscle Research

*The contractile tissue has many remarkable characteristics. Not the least of them is its usefulness in the investigation of life itself.*

## A. Szent-Gyorgyi

When one of our muscles is excited by a nerve, it contracts. Some muscles are thus able to pull at the bones to which they are attached and to move a part of the body. Other muscles perform much of the work of our internal organs. The heart is only a pouch of muscle; so, in a way, are the intestines. Millions of muscle cells are embedded in the walls of small arteries.

Much of human suffering is due to the dysfunction of these inner muscles. A slight contraction of the muscles of the arteries may send the blood pressure up or cut off the supply of oxygen to other tissues. More than half the deaths of mankind are due to the failure of heart muscle. Yet in most cases we cannot repair the diseased muscle machine because we do not sufficiently understand its structure and function.

Muscular contraction is one of the most wonderful phenomena of the biological kingdom. That a soft jelly should suddenly become hard, change its shape and lift a thousand times its own weight, and that it should be able to do so several hundred times a second, is little short of miraculous. Undoubtedly muscle is one of the most remarkable items in nature's curiosity shop. Muscle, however, has attracted the attention of many scientific investigators for another reason.

All living organisms are but leaves on the same tree of life. The various functions of plants and animals and their specialized organs are manifestations of the same living matter. This adapts itself to different jobs and circumstances, but operates on the same basic principles. Muscle contraction is only one of these adaptations. In principle it would not matter whether we studied nerve, kidney or muscle to understand the basic principles of life. In practice, however, it matters a great deal.

The work of the scientist is essentially to measure, and the rapid changes in muscle can be measured much more easily than the slow changes in liver or kidney. The functioning of muscle may be seen by the naked eye, and may be indicated by simple means. The electrical change of nerve, on the other hand, may be observed only with involved and subtle devices. The great motility of muscle demands that it be built of small units, arranged with great regularity and bound together by relatively weak forces. This means that we may disentangle and isolate these small units without destroying them, and that they may be studied outside the body.

There are many approaches to the study of muscle. The anatomist delights in its structure, which he tries to preserve. The physiologist enjoys the harmony of its function, and tries to avoid all damage to the tissue in order to study it under physiological conditions. The biochemist, however, willfully destroys the structure. There is a simple reason for the fact that none of these approaches, in itself, can explain muscle.

Muscle is a machine, and in any machine we must deal with two elements. One is the energy-yielding reaction, such as the expansion of steam in a steam engine, the burning of fuel in an internal combustion engine, or the flow of current in an electric motor. These elementary reactions can accomplish useful work only if they take place within a specific structure, be it a cylinder and a piston or a coil and a rotor. So in muscle we must also look for both the energy-yielding reaction and the meaningful structure.

The energy-yielding reaction is a chemical change which takes place among molecules, and its study belongs to the realm of biochemistry. The structure is the domain of the anatomist, working with his knife, microscope or electron microscope. Both paths of inquiry are most exciting. We can expect to find that the basic energy-yielding reaction is identical, at least in principle, in all living forms. Muscle research can thus take us to the very foundation of life. Its structure, although specialized, can likewise reveal the fundamental principles of biomolecular architecture. In this light muscle ceases to be a special problem. The study of its function merges with the study of all life, and for such study muscle is a wonderful and unique material.

**M**uscle is built of tiny fibers which are just visible to the naked eye. If we were to look at a picture of these fibers closely, we would discover a horizontal striation which indicates that the muscle fiber is only a bundle of many small fibers, or fibrils. These fibrils are composed of contractile matter, and the muscle contracts because they contract. Under the electron microscope the fibril is clearly visible as a slender, continuous column.

The fibrils need a great deal of energy in a very special form. The energy contained in food, in fact, must be converted into this form before the fibrils can use it. This process alone requires a bulky chemical apparatus that is located between the fibrils. The contraction of the fibrils presses this substance into disks, which account for the vertical striations The inter-fibrillary material still adheres to the surface of the fibril, giving it a segmented appearance. Where the fibril itself lies naked we can see that it too is a bundle of still finer threads. These threads have been called filaments by C. A. Hall, M. A. Jacus and F. O. Schmitt, who first photographed them at the Massachusetts Institute of Technology.

The specific form in which energy is supplied to the fibril is adenosine triphosphate, which is abbreviated as ATP. The discovery of this substance is one of the most important achievements of biochemistry. ATP is one of the main axes about which life revolves. The ATP molecule bears three phosphate groups linked by oxygen atoms. The University of Pennsylvania physiologist O. Meyerhof has shown that the manufacture of each such link requires 11,000 calories of free energy. When the links are broken, the energy is released. Fritz Lipmann of the Massachusetts General Hospital has called them high-energy phosphate links. Their splitting is the source of all muscular energy.

The participation of ATP in contraction has one most fascinating aspect. Our experience thus far indicates that wherever there is life its carrier is a nucleo-protein. This substance is made up of protein and nucleic acid, the latter being most abundant in cell nuclei. We may therefore assume that these two materials and their interactions are an essential feature of life. The nucleic acid molecule is composed of many small units that are chemically similar to ATP. These are joined in giant fibrous molecules. Unfortunately we cannot do much with such large molecules, so they can tell us little of the nature and meaning of nucleoproteins.

Again muscle furnishes a hopeful exception. So far as the author is aware, the contractile substance of the fibrils is the only "living" protein that is not linked to nucleic acid. The reason is easy to find. Long, fibrous nucleic acid molecules would surely interfere with the mechanism of contraction. Instead of nucleic acid, the contractile protein works with smaller units. In the author's opinion, ATP is the missing nucleic acid of the fibril. Since ATP is a small molecule, its connection with muscle protein can be studied with relative ease, and may reveal one of life's most closely guarded secrets.

Having outlined some general principles, let us see what we can actually learn about the function of muscle. To understand it we will have to break it down. From a knowledge of its parts we may hope to understand the whole, just as the astronomer understands the stars by his knowledge of the atom. A man who parachuted into a strange land, however, might have difficulty in finding his way home. We would likewise have trouble finding our way back to muscle if we decomposed it all at once. We will fare better if we proceed to our destination on foot, and decompose muscle step by step.

As the first step in decomposing muscle, let us separate the small, soluble molecules from the insoluble structure without destroying the latter.

This can easily be done by washing a strip of rabbit muscle, for example, with water. Now let us put the dissolved molecules back again. We suspend our muscle in a concentrate of the original solution, or, to simplify matters, in a boiled extract of some other muscle tissue. What happens? The muscle contracts! Special measurements show that the contraction occurred with such force that the muscle could have lifted a thousand times its own weight, just as any living muscle. There is no doubt that what we have seen is muscle contraction.

Next we ask: What substance in this muscle juice made the washed muscle contract? Fortunately this is not a difficult question. A bit of scientific cookery indicates that two substances were responsible—potassium and ATP. This is remarkable, for ATP is essentially phosphate, and if we want a lawn to flourish, we use a fertilizer containing potassium and phosphate. Grass needs the same substances as muscle, a striking demonstration that the basic change in muscle contraction is only one form of a universal biochemical reaction.

We have thus discovered that ATP makes muscle contract as well as supplying the energy for contraction. No other substance will serve. ATP is a cogwheel in the mechanism of contraction, and without it no contraction occurs. ATP has yet another function. From earliest times man has known of *rigor mortis,* the stiffening of the body that follows death. The same effect can be produced in a single muscle. A strip of rabbit muscle removed soon after killing the animal can be stretched like rubber, though within narrower limits. A few hours later the strip becomes inelastic and simply breaks when we attempt to stretch it. The loss of elasticity is due to the decomposition of ATP; by restoring ATP we restore elasticity. If ATP did not make muscle elastic, the muscle would be too rigid to work at all.

We can now go one step further in decomposing muscle. We make a very fine pulp of our washed rabbit muscle, suspend it in water and add potassium and ATP. No contraction can occur because the structures have been destroyed. Instead there is a violent precipitation. Much the same thing must have happened when the whole muscle contracted. The precipitation of fine colloidal particles is due chiefly to a loss of electric charge. So the basic reaction of contraction is a loss of charge, brought about by potassium and ATP.

We may now proceed to decompose muscle into its molecules. In the presence of ATP a strong salt solution dissolves the muscle structure and extracts two quite different proteins—actin and myosin. Both of these proteins have most interesting properties, but contractility is not one of them. The most amazing property of myosin is its great affinity with ions, which in the smallest concentrations may greatly modify its electric charge. The most amazing property of actin, discovered by the author's associate F. B. Straub, is that it can exist in two forms. When first extracted it consists of small, round molecules. If we add a little salt these little globules unite to form long threads, as shown in the photograph at the bottom of the opposite page.

The most amazing property of actin and myosin, however, is that they can unite to form a complex—actomyosin. It is this complex that has the contractility of muscle. We have reproduced, and thus made it possible to analyze, one of the most mysterious manifestations of life. Seeing actomyosin contract for the first time was the most exciting experience of the author's scientific career.

We have left the problem of the molecular architecture of muscle to the end, because this field of research was opened not long ago when the discovery of the electron microscope extended the limit of human vision down to the world of molecules. Hall, Jacus and Schmitt were the first to begin studies along this line. At present the problem is also being investigated at the National Institute of Health by G. Rozsa, the author and R. W. G. Wyckoff, in the laboratory of the latter. These studies have shown that the building of actin out of globules into fibers proceeds in two steps. First perhaps 20 globules unite into a slightly elongated particle some 300 Angstrom units long and 100 wide. Then these particles are joined end to end to form threads. Electron-microscope photographs show that the threads have a strong tendency to come to rest side by side so that the individual particles of neighboring threads also lie side by side. Thus threads actually form in two directions, and a regular structure analogous to a crystal results.

The nature of the chemical mechanism that makes large protein molecules out of small ones is a very fascinating problem. The larger actin particles are the size of the smaller viruses. The question of how a large protein unit is put together, and how it is taken apart and put together again according to a new pattern, is perhaps the most important problem of virus research. W. J. Bowen and K. Laki of the National Institute of Health have shown that ATP is also involved in this feat of molecular engineering.

There is yet another new and rather hopeful approach to muscle, opened by the discovery that muscular contraction is a so-called equilibrium reaction. The approach is that of thermodynamics. It has thus far yielded two rather fascinating results. It has shown that the whole muscle machine is built of small and more or less independent units the size of actin globules. These have a molecular weight of 70,000, *i.e.,* 70,000 times the weight of the hydrogen atom. If linked to ATP, the units at rest have a certain amount of potential energy, and are thus comparable to a loaded gun or an extended spring. When excited by a nervous impulse, the units spend this energy. The energy transmitted by ATP is then required to bring them back to the high-energy loaded or extended state. Once we know this, it seems natural that nature should do it this way. The living structure is kept ready to fire and is loaded after firing, instead of the other way around.

Can we put all this together in a single theory of muscle contraction? We can try, but we must jump some rather wide gaps. We have seen that the primary reaction is a change in which electric charge is lost, and that the change takes place within small molecular units. Such a loss of charge must very greatly alter the forces between the larger particles. Should it be found that the muscle fibril is a three-dimensional structure of slightly elongated particles, it might also be found that contraction is nothing more than a tilting of them.

Much work must be done before we have a rounded understanding of muscle contraction itself. Then we must ask what changes occur when a nerve commands a muscle to contract, how the system returns to its resting state, and how the energy of ATP is transferred. Since actomyosin appears to be identical in all kinds of muscle, we will have to look out for the substances that regulate its varying functions. In the wing muscle of some insects, for example, actomyosin can contract several hundred times a second, yet it can also produce the slow, regular beat of the heart and, without consuming energy, can hold a clam shell closed for hours.

The reader may ask: When we know all this, will we understand muscle and life itself? The author can only give his personal opinion: We will not, because the fundamental changes in muscle cannot be expressed in terms of orthodox chemistry. We will very likely have to explain them in terms of how electrons are distributed over the entire molecular structure, an explanation that belongs to the realm of quantum mechanics.

The study of this distribution of electrons within the protein molecule is one of the most urgent and most difficult tasks of biology. Until the task has been completed, we cannot hope to understand the nature of life. The task is not impossible, but it requires keen imagination, a lust for adventure, and a catholic knowledge. The task is probably too big for any one man; the biologist and the theoretical physicist will have to collaborate. A few hopeful beginnings have been made. They may eventually lead us to a full understanding of the protein molecule, which will mark the beginning of a new era in biology and medicine.

## ABOUT THE AUTHOR

**A. Szent-Györgyi** (1893-1986) was awarded the Nobel Prize in Physiology or Medicine in 1937 for his discoveries in connection with the biological combustion processes, with special reference to vitamin C and the catalysis of fumaric acid.

# The Influenza Virus

*The organism which causes the disease that sweeps nations is curiously changeable. This makes it difficult to anticipate epidemics with vaccines, and suggests a future hazard to man.*

## Sir Macfarlane Burnet

In a world from which medical science and social progress have almost eliminated serious infectious disease, influenza holds a unique position. It is the only acute infectious illness that still attacks most adults in advanced Western countries. And it is the only disease which in the 20th century has shown a capacity to increase in virulence and sweep unchecked around the world, killing millions in the process. That happened in 1918-19. Could such an outbreak occur again, and if it did, could our present-day vaccines and drugs control it? No one can yet say.

The main objective of work on influenza is to be prepared to deal with a lethal pandemic of that type, should one appear again. The periodic epidemics of influenza that we have had in recent years, one of which is now running its course in North America and Europe, have been of a milder form. As long as influenza retains the character it has shown since 1933, our practical objectives are limited. Immunization with vaccines has been applied mainly to military personnel for the purpose of avoiding loss of time, particularly during training. Theoretically it might be desirable to protect the whole population by similar methods; what is good for the soldier surely is good for the civilian. Actually not many people regard influenza as a serious enough danger to want to have one or two immunizing "shots" each fall, and without a better knowledge than we have of what strains of influenza viruses are likely to produce epidemics, it is difficult to become enthusiastic about large-scale immunization.

Yet there is one group in the civilian population that needs close study, with a view to deciding whether we should try to immunize all its members each year. These are the elderly people. Whenever an epidemic of influenza passes through a community, there is a sharp peak of deaths from various causes among the aged. Any elderly person rendered frail by physical disability is likely to succumb to an attack of influenza. This was heavily underlined during the 1951 influenza outbreak in Great Britain. In Liverpool the epidemic passed like an angel of death amongst the old. During the peak week there were more deaths than in the worst week of the 1918-19 pandemic. An investigation of the saving of life that might be effected by appropriate immunization of the aged against influenza would seem to be a very worthwhile project.

We now have a substantial fund of knowledge about the virus of human influenza, which was first isolated in 1933. For this knowledge we are largely indebted to two major technical discoveries made about 1941: (1) a convenient way of growing the virus (in chick embryos), and (2) a test to determine the presence of the virus (the clumping of red blood cells suspended in a test tube). Thanks to these laboratory aids, it is hardly too much to say that the influenza virus today is as well understood as the diphtheria bacillus or the pneumococcus. They have given a tremendous stimulus to the investigation of all aspects of influenza.

The first requirement in the study of any disease is to be able to define it in terms of the responsible microorganism. This is especially important in the case of influenza because its symptoms are just like those of at least a dozen other known infections, notably certain common feverish colds which are quite indistinguishable clinically from mild influenza.

One way to identify influenza is to isolate the virus. This is normally done by inoculating into chick embryos throat washings taken from persons in the acute febrile phase of the illness. After four or five days' incubation, the embryo fluid will contain enough virus to agglutinate red cells. Thereafter the virus can be transferred to fresh embryos for further study.

A more convenient way is to take two samples of blood from the patient, one at the earliest stage of the illness, the other two weeks later, and test the serum for the presence or absence of antibody against standard types of virus. If in the second blood specimen five out of six patients of the sampled population have antibody against a specific influenza virus, we can be certain that the outbreak was due to that type of virus. If, on the other hand, none shows an increase in antibody against any type of influenza virus, we must look for another cause of the epidemic.

This use of immune reactions for diagnosis introduces us to the important question of the different species and types of influenza virus. There are three species, called influenza A, B and C. Immunologically the three are completely distinct. Infection with one produces no antibodies or immunity against either of the others. Influenza C is a rare type which apparently produces only trivial illness; it need not be further considered. Of the other two, influenza A occurs more frequently and in larger and more severe epidemics than influenza B. Most of the research has been done on influenza A.

An influenza A virus is defined as a virus which reacts with antibody produced by infection with a standard A-type virus. But A viruses differ among themselves in various ways. For instance, on the basis of immunity tests in animals we can identify three distinct strains, called A1, A2 and A3. All three cross-react in the test tube, but in experimental infections they induce very little immunity against one another. There is good evidence from large-scale tests in the U.S. Army that a vaccine prepared from A2 virus was effective against an A2 epidemic in November, 1943, but was quite useless against an A3 epidemic which reached North America in February, 1947.

The difference is one of degree, and the most convenient way of studying it is by what we call hemagglutinin inhibition tests. Given a set of the three A-type strains and a set of the three immune blood sera obtained from animals infected with the same three strains, we perform a series of cross-reactions to determine the least amount of each serum that will prevent a standard dose of each virus strain from agglutinating chicken red cells. It turns out that while the A1 antiserum, for example, can inhibit the effects of all three viruses, far less of it is required to counteract the A1 virus than either of the others. A1 antiserum from a rabbit can be diluted to one part in 2,000 and it will still neutralize the A1 virus, but to inhibit the A2 virus the serum must have a concentration of one part in 200, and to counteract the A3 virus, one part in 100. Similarly, A2 virus is effectively counteracted by A2 antiserum in a dilution of one part in 3,000, but the same serum must be concentrated to one part in 400 to inhibit the A1 virus, and one part in 300 to inhibit A3. In practice this means that A1 antibody is ineffective against the A2 virus, say, because it is not usually present in high enough concentration to neutralize the latter.

With this background we can turn to the history of influenza from the time when it became possible to study the virus in the laboratory. Influenza A has recurred in the Northern Hemisphere every second or third year. Some outbreaks have been more widespread and severe than others. In England, for instance, 1933, 1937, 1947 and 1951 were years of severe epidemics. The activity of influenza B is less clearly defined: the U.S. suffered outbreaks of the B type in 1940 and 1946.

One simple generalization emerges from the work of the last 10 years: the virus changes type frequently. When it does so, the new type rapidly supersedes its predecessor. The serological character of the A virus has changed seven or eight times since 1933, and each change in character has within a year been evident all over the earth. Soon after influenza A2 was found in the U.S., it appeared in Australia and England as well. After it had taken hold, no A1 strains were found anywhere. And so for each successive change. Occasionally an anomalous virus that seems out of step with the current type does appear, but by and large the rule of world-wide dominance by one type holds. This is a phenomenon of great interest to epidemiologists: it may well provide the key to the understanding of influenza.

There are diseases that maintain their character unchanged for centuries. The mumps virus, for instance, apparently caused much the same symptoms in Hippocrates' time, 2,500 years ago, as it does today, and we can find only trivial immunological differences among present strains of the virus. One attack of mumps will give practically lifelong immunity to the disease. Influenza, on the other hand, gives rise to an immunity which becomes ineffective as soon as a new strain of the virus appears. It also possesses another peculiar feature. In every influenza epidemic many people who never "report sick" are infected mildly with the virus and develop immunity. After a widespread epidemic has passed through a city, few of its people will lack antibody against that particular type of virus.

Here we have a key to the behavior of the influenza A virus. It is a parasite whose only natural host is man. To survive, it must pass continually from one human being to another, but from the very effectiveness of its means of transfer—it is inhaled and lodges in the respiratory tract—it soon finds itself in the position epidemiologists call "exhaustion of susceptible hosts." In other words, almost the entire population becomes immune. In such circumstances there are two ways by which a highly transmissible virus can survive. It may, like mumps or measles, live at the expense of the newborn and become a children's disease. Or it may meet the situation, as the influenza virus does, by a transformation of character—a mutation that enables it to overcome its host's immunity. The importance of such mutations is emphasized by their swift race over the world. In 1946 a new type of A virus was isolated in Melbourne. Australia is a long way from the great centers of population in the Northern Hemisphere. But within a year the new virus was causing widespread influenza in those distant places. Influenza vaccines from previous types of virus were powerless against that sudden epidemic of February, 1947. (Their failure, incidentally, was largely responsible for our first serious recognition of the importance of the influenza viruses' continual change in type.) From 1947 to 1950 the Melbourne virus was the only type found world-wide.

Immunity is not the only quality that changes. The virus probably maintains a continuous series of mutations, each producing inheritable changes in this or that quality. Each new form must, as it were, go through the test of survival in competition with the old form. It may take a variable time before a really successful competitor emerges. But when it does appear, its success is apt to be so overwhelming that a wave of influenza ripples over the whale of the world. The epidemics may not be severe, but in one form or another each wave represents a new pandemic. It is reasonable to believe that the great pandemic of 1918–19 was the result of the emergence of a series of influenza virus-A mutations which eventually developed a virulence and invasiveness beyond any farmer precedent.

From what has been said about the changing immunological character of influenza viruses, it will be realized that to provide effective immunization we must make a vaccine from virus types as close as possible to the one that will be dominant during the coming winter. The main hope of doing this is to have in existence an organization for watching the immunological changes of the virus in all parts of the world. If, far instance, a winter epidemic in Australia is found to be due to a new type, that particular type should certainly be incorporated in any North American vaccine far use during the following winter. If we should have a repetition of 1918-19, there would be a special need far swift recognition of new types and streamlined progress from the laboratory to mass production of the vaccines. The World Health Organization's Influenza Center in London is the laboratory primarily concerned with keeping in touch with the changing pattern of influenza viruses. There is a second center in the U.S. and a large number of collaborating laboratories around the world.

In one sense everything we have discussed so far is at a rather superficial level. To go a little deeper into the problem of the changes in viruses we must enter the newest of the biological subsciences, virus genetics.

Everyone who has handled influenza viruses in the laboratory has experienced their capacity to undergo inheritable changes in character—to mutate. In fact, every time we isolate virus from a lmman throat we automatically sort out those mutants that multiply readily in the chick embryo. The cells of the chick embryo favor certain mutant types different from the forms favored by the cells of human air passages. By now there is a great deal an record about the types of variants that can be selected out in the laboratory, and it is becoming clear that an influenza virus is far from being as simple as its small size and undistinguished appearance under the electron microscope might suggest.

Within the last three years a new approach has been made in my laboratory toward a further analysis of variation in influenza viruses. The technique is basically similar to hybridization in higher organisms. It is well known that when two bacterial viruses infect a single bacterium, they may give rise to progeny with qualities derived from both viruses; we might almost call them parents. We have now found that influenza viruses, like the bacteriophages, also interact genetically. When two different influenza virus particles not too distantly related to each other can be induced to infect a single cell at the same time, more than two types of descendent virus eventually emerge. The genetic character" of the parents are reshuffled and recombined in them. It is hard to believe that anything really comparable to sexual fusion takes place, but there is only a rather vague and speculative hypothesis to offer as an alternative.

If we try to combine the results from virus work with what is known to date of influenza virus multiplication within the cell, the picture that emerges is something like this. A virus can be regarded as composed of two systems. One provides the means of entry into the host cell and incidentally carries the immunological qualities that are so important for medicine. These are essentially surface qualities and can be regarded as equivalent to the body tissues of a higher organism. The second system is probably made up of genes and is responsible for the inheritable character of the virus. Apparently the only function of the first system—the somatic portion of the virus surface—is to make possible the virus entry into the cell. Once the virus particle has gained entry, its soma seems to play no further part. Only the genetic material is concerned in the next stage. In one way or another the genes manage to draw from the host cell what they need to duplicate and reduplicate themselves, until there are about a hundred copies of each gene in the infected cell. It is not yet clear whether the whole group of genes duplicates as a unit—one might picture it in the cell as a thread of genes splitting into two threads—or whether each gene goes about its duplication in partial or complete independence of the others. If the first is the case, the genetic interaction between two types must be analogous to "crossing over" in classical genetics. In either case the process must eventually reach a stage where many complete new gene groupings are available. Each of these must then in some way control the construction of a new soma around itself. Until this is done, no detectable virus particle can emerge from or be artificially extracted from the infected cell. Clearly the mystery of biological replication is no less a mystery in the smallest of living organisms than in the largest.

Whatever the mechanism, this process of recombining qualities from two different strains provides us with a powerful method of analyzing virus activities. The genetic approach is of extreme importance for every type of biological problem. Biochemistry and bacteriology in recent years have been completely rejuvenated by its application, and it can hardly fail to yield similar help to virus research. There seems to be no theoretical reason why we should not eventually obtain a detailed understanding at the genetic level of the nature of virulence, and of the nature and limitations of immunological changes in influenza viruses. That might still leave us a long way from understanding these qualities in physical-chemical terms, but probably it would provide all that would be needed at the important clinical level. We could look forward to being able to predict the changes in influenza epidemics from year to year and to preparing appropriate vaccines. The same approach might also prove applicable to other viruses and cut short the work needed to find the best types of immunizing agents for various virus diseases, such as poliomyelitis, for example.

There is, however, one thing which we dare not forget. In dealing with viruses responsible for human disease, we are dealing with matters of life and death. Full knowledge of the influenza virus would enable us to save life, but it would also make it possible to let loose, maliciously or by accident, another plague like that of 1918–19, which spread unchecked through civilized and uncivilized countries alike, and in the process killed

25 million people. Microbiology is bound to advance just as nuclear physics has advanced, but perhaps even more manifestly the impact of that advance on human affairs will depend upon the way in which it is applied. Medical microbiology in the past has produced the most directly beneficial of all social revolutions by removing most of the dangers of infectious disease. It would be a bitter irony if its further development should bring man-made plagues even more lethal than the natural epidemics of the past.

## ABOUT THE AUTHOR

**Sir Frank Macfarlane Burnet** (1899-1985) was awarded the Nobel Prize in Physiology or Medicine in 1960 along with Peter B. Medawar for their discovery of acquired immunological tolerance.

# The Life Cycle of a Virus

*What happens in the time between the infection of a cell and the appearance of new viruses? The behavior of bacteriophages suggests that this stage involves a new concept: the provirus.*

## André Lwoff

*Close your eyes and look. What you saw at first is there no more; and what you will see next has not yet come to* —Leonardo da Vinci

We can apply these words very aptly to a virus—of the bacteria-infecting kind known as bacteriophage. When a phage particle enters a cell, it loses its infective power and its identity as a particle. Generally its entrance into the cell is followed within 15 minutes to an hour by the emergence of a new generation of infectious virus particles. Sometimes, however, there is no immediate pathological event. The genetic material of the virus that has passed into the cell combines with the genetic material of the cell itself. In doing so it is converted into something that has been named a "provirus," meaning before virus. Days or years afterward the provirus may suddenly develop into virus and the bacterium give rise to a group of virus particles.

The term provirus needs some explanation. The expression "proman" would certainly not evoke the idea of a human egg, from which *Homo sapiens* always develops, but rather that of an evolutionary ancestor of man which would have to undergo a genetic transformation to become man. A provirus may perhaps correspond to an evolutionary ancestor of a virus. But it is also much more than that.

Before attacking the question of the nature of proviruses, we must know something about viruses themselves. What is a virus? We shall leave out of the discussion the much debated issue as to whether viruses are living organisms or not; our concern is to find out how they differ from "normal" organisms of the microbiologist's world. The two attributes that are usually thought to define viruses are their very small size and the fact that they can multiply only inside living cells—usually requiring a specific kind of cell host. But to learn more about their peculiarities let us go beyond this definition and compare viruses with other small biological units.

First of all, how does a virus differ from a cell? Most cells are capable of reproducing themselves: they possess the genetic material which is the basis of heredity and the tools necessary to synthesize the essential building blocks and to organize these into a structure just like themselves. We can see three important differences between a cell and a virus, taking bacteriophage as a typical virus: (1) whereas cells contain both desoxyribonucleic acid (DNA) and ribo-nucleic acid (RNA), the phage contains only DNA; (2) whereas cells are reproduced from essentially all their constituents, bacteriophage is reproduced from its nucleic acid; (3) whereas cells are able to grow and to divide, the virus particle as such is unable to grow or to undergo fission. Bacterial viruses are never produced directly by division of all existing virus; invariably they are formed by organization of material produced in the host cell.

Next we must consider whether viruses bear any likeness to the particles within a cell, particularly the particles called plasmagenes. Here differences are less easy to find. The theory has been proposed that viruses may originate as mutated plasmagenes. But we know that some plasmagenes (*e.g.,* the chloroplasts of green plants) can grow and divide. Furthermore, plasmagenes are not pathogenic or lethal to the cell, as virus particles are. Let us just note, for the time being, that nothing which resembles a bacteriophage in its properties, life cycle, shape or organization has been found in normal cells.

Let us now consider the peculiar behavior of the virus. The virus particle as such is only the beginning and the end of a life cycle. Its only physiological function is to obtain entry of the virus' genetic material into the host cell. After this occurs, what remains of the virus is devoid of infectious power. There follows a vegetative phase in which the specific constituents for new viruses are produced. Finally these constituents are organized into virus particles, which are liberated by lysis (dissolution) of the cell. The whole process usually takes from 15 to 60 minutes.

But a bacterial virus may multiply in another way, and this is where the pro-virus enters the picture. Ordinarily the virus nucleic acid passed into a cell proceeds promptly to multiply and to synthesize specific protein material for new phage particles. Sometimes, however, the nucleic acid may anchor onto a bacterial chromosome and act as if it were a normal constituent of the cell. It behaves as a bacterial gene, being replicated at each bacterial division and transmitted to each daughter bacterium. This is what we call a provirus. It is a potential virus; it may eventually give rise to virus particles. In the meantime the bacterial offspring go on growing and dividing as normal bacteria, and each daughter bacterium yields progeny capable of producing viruses. In other words, the ability to produce viruses is perpetuated inside the bacterium; no new infection from outside is needed.

Bacteria containing proviruses are called lysogenic. When a small number of such bacteria are broken down, no infectious particles can be found. This means that the provirus is not infectious. And yet in every large population of these bacteria some mature bacteriophage particles appear. From time to time a bacterium in such a culture suddenly disappears, and about 100 phage particles emerge. The probability of a lysogenic bacterium spontaneously giving rise to viruses varies from 1/100 to 1/100,000. In some systems the probability is apparently independent of external factors; it cannot be modified. In other lysogenic systems phage production may be initiated at will by inducing agents, such as X-rays, ultraviolet rays, nitrogen mustard and other substances-all of which are known to be capable of producing mutation. Within 30 to 60 minutes after exposure to one of these agents, practically all of the bacteria produce viruses and lyse.

How do lysogenic bacteria produce viruses? Before discussing this question we must know more about the proviruses. We are inclined to think that pro-viruses originally arose as mutants of normal bacterial genes. Whatever their origin, the reservoir of bacterial viruses seems to be the provirus-carrying bacteria. These bacteria may have perpetuated provirus, that is to say, the hereditary ability to produce virus, for many thousands of years.

The study of lysogenic bacteria has led to a clear picture of the provirus. Apparently it does not contain virus protein, for lysogenic bacteria do not cause the production of specific antibodies to phage protein in experimental animals. It is therefore tempting to visualize the provirus as a large molecule of nucleic acid. Secondly, the provirus is associated with a certain genetic character of bacteria and is located at a specific site on a bacterial chromosome. Thirdly, two genetically related proviruses in a bacterium may cross over and recombine. Fourth, a lysogenic bacterium is immune to infection by a phage particle related genetically to its provirus, though it can be killed by an unrelated phage. As long as the provirus remains in that state, a genetically related superinfecting phage is unable to develop into phage. Finally, the mere presence of the provirus may modify the properties of a bacterium: it may endow certain bacteria with the ability to produce a toxin they could not otherwise make, or it may change the typical appearance of bacterial colonies. Things happen as if the provirus either carries a specific gene or modifies the neighboring bacterial genes. From all these data it may be concluded that pro-virus is the bacterial virus' genetic material, bound to a specific site in the bacterium and responsible for a specific bacterial immunity.

Now it is difficult to imagine that this immunity is due only to the *presence* of the provirus. A particle cannot exert a specific action by its mere presence. The only way the provirus can make the bacterium immune—that is, prevent multiplication of a virus invader—is by modifying or blocking a specific activity of the bacterium necessary for that reproduction. And the provirus can do this only if it is present at a specific site. As a matter of fact, we can account for all the properties of proviruses and of lysogenic bacteria by the hypothesis that the provirus is the genetic material of the virus anchored at a given site in the bacterium. The genetic material of an infecting virus becomes a provirus when and because it becomes bound at that site to a specific receptor, which modifies the material. It then gives the bacterium immunity against genetically related infecting particles. An inducing agent such as ultraviolet rays destroys the immunity because it displaces the genetic material of the virus from its specific site.

For a long time virologists have concentrated on the virus particle itself. Yet the particle is only a prelude to the infection. During the longest and most important part of the life cycle, the pathogenic phase in a cell, no virus particle is present. As a matter of fact, disappearance of the virus particle is the *sine qua non* for the development of the cellular lesion. Indeed, there are cases in which all the bacteria in a lysogenic population die although very few of them produce bacteriophage particles; the cells are killed by a defective development of proviruses initiated by an inducing agent. One could even conceive of a condition in which the probability of the virus ever appearing would be infinitely small, that is to say, practically absent. Some bacteria actually carry a gene which can initiate the synthesis of a protein lethal to themselves. But that is another story.

Biologists have long been accustomed to think of death in terms of the destruction or alteration of some vital structure. We have been less inclined to think of living cells as carrying the seeds of their own destruction, or of the possibility that lethal agents may kill in more than one way. For example, X-rays sometimes kill by destroying essential structures, but they may also destroy a cell by inducing a gene to express its lethal potentiality. This potentiality is sometimes the power to start a new synthesis which may or may not end in virus particles.

To what extent are the phenomena disclosed in bacteria valid for higher organisms? May animal or plant cells perpetuate proviruses? Are some viral diseases of man the result of the activation of a provirus? May immunity to virus diseases be explained in terms of proviruses? Do the findings concerning lysogeny have any bearing on cancer? Let us recall that the inducing agents which can trigger proviruses to give rise to viruses are all not only mutagenic but also carcinogenic—radiations, nitrogen mustard and so on. It is indeed tempting to theorize that carcinogens may induce malignancy by initiating the formation of a pathological structure from a provirus-like material. Many facts are in favor of the hypothesis that proviruses originate some animal diseases, but the problem cannot be discussed within the limits of this article. Suffice it to say that this is, at any rate, a working hypothesis.

I have tried to outline the concept of the provirus, to analyze its relations with the concept of the cell and of the virus and to show the impact of the newly acquired knowledge on our conceptions of cellular disease. The common denominator of the various phases of the life cycle of a virus is the genetic material— the nucleic acid—which may exist in three states: infectious, proviral and vegetative. Throughout these three states the genetic material apparently remains essentially the same in structure, but it changes radically in dynamic potentialities and behavior. The virus particle, the end product of the vegetative phase, is a quiescent nucleoprotein particle, unable to grow or to divide. The provirus is an integrated nucleic acid, which behaves like a gene and is replicated like the host genes. Neither the virus particle nor the provirus is pathogenic *per se*; their pathogenic property is only potential. The only pathogenic phase of the virus is the vegetative phase, during which the specific viral nucleic acid multiplies and during which the specific viral protein is synthesized. Things happen as if the synthesis of the protein is responsible for pathogenicity.

The provirus produces provirus; it is order. The vegetative particle produces virus particles and a disease of the host; it is disorder. The virus particle does not produce anything; it is an extremely conservative particle—the absence of any activity, that is to say, a kind of order. Thus the virus is an alternation of order and disorder.

As a result, my presentation of the subject may seem somewhat disordered. For this I had decided to apologize, when I came across an unpublished letter which Martin de Barcos, Abbot of Saint-Cyran, wrote to Mother Angélique in 1652: "Allow me to tell you that you would be wrong to apologize for the disorder of your discourse and of your thoughts, because, if they were otherwise, things would not be in order, especially for a person belonging to your profession. As there is a wisdom which is folly before God, there is also an order which is disorder, and in consequence, there is a folly which is wisdom and a disorder which is the true rule." This being exactly the case of the virus, I decided not to apologize.

## ABOUT THE AUTHOR

**André Lwoff** (1902-1994) was awarded the Nobel Prize in Physiology or Medicine in 1965 along with François Jacob and Jacques Monod for their discoveries concerning genetic control of enzyme and virus synthesis. Lwoff discovered the inductive action of ultraviolet irradiation.

# The T2 Mystery

*T2 is a virus which dissolves bacteria.*
*Normally its attack is followed by the appearance*
*of a generation of new viruses. But sometimes*
*the viruses appear to be missing. Why?*

## Salvador E. Luria

Our story has as its critical episode one of those coincidences that show how discovery often depends on chance, or rather on what has been called "serendipity"—the chance observation falling on a receptive eye. The episode is a good illustration of the principle of "controlled sloppiness," which states that it often pays to do somewhat untidy experiments, provided one is aware of the element of untidiness. In this way unexpected results, sometimes real discoveries, have a chance to come up. When they do, we can trace their cause to the untidy, but known, features of the experiment.

The story has to do with bacteriophages, or bacterial viruses. The habits and reproductive cycle of these bacteria-infecting viruses are familiar to the readers of SCIENTIFIC AMERICAN. A virus particle attaches itself to a susceptible bacterium and injects its reproductive material, mainly nucleic acid; this material multiplies in the bacterial cell, and within half an hour the bacterium dissolves and out come hundreds of new mature virus particles.

In 1946, while experimenting with infection of the common colon bacterium *Escherichia coli* by the bacterial virus called T2, I noticed a peculiar violation of the usual pattern of events. Certain mutant strains of the bacterium took up the virus, were duly dissolved after the customary period but produced no detectable viruses! When the material was tested, no trace of infectious virus could be found in it. I explored this phenomenon a little further, but after playing around with it for a few weeks and getting nowhere, I shelved it in my mental files as the "T2 mystery."

In 1950 I returned to the problem. I had become interested in the study of incomplete virus particles as possible precursors of viruses, and it seemed that the juice from the bacterial mutants might be a good place to look for such precursors—arrested viruses, as it were. I proceeded to re-examine the matter with a co-worker, Mary Human.

One day, in preparation for more complicated experiments, we decided to add some streptomycin to the juice from the dissolved bacteria. To carry out the measurements we planned to make, we needed bacteria resistant to streptomycin. It happened that no streptomycin-resistant culture of *Escherichia coli* had been prepared in the laboratory that day. Rather than wait, Mrs. Human decided to use an available streptomycin-resistant culture of another bacterium which is susceptible to T2: namely, the dysentery bacillus (*Shigella dysenteriae*). Of course the substitution made it not a "clean" test. But since virus T2 seemed to behave alike on both hosts, it hardly seemed to matter.

The next day the T2 mystery was solved; or rather, as often happens in science, it had been transformed into a bigger one. The juice from the dissolved coli bacteria, which had seemed virus-free, raised havoc with the dysentery bacilli. In other words, it contained plenty of infectious virus, but the virus was infectious only to the dysentery bacteria not to the coli. The mutant coli cells in which the virus had reproduced had changed it somehow. But the change was not profound: we discovered immediately that after a single cycle of reproduction in the dysentery bacilli, the virus reverted to the original T2 type—that is, it could infect coli again!

This was a great surprise. If the virus had undergone a stable, hereditary change during reproduction in the unusual, mutant coli, that would have been understandable. It is not uncommon, when a virus invades a new host, for a mutant type of virus to emerge and become dominant. In that case the host has simply favored mutant viruses which happen to be present; it has not itself modified the virus. But no mutation was involved in the change of the T2 virus to the new type and back. Every T2 particle multiplying in mutant coli produced only progeny of the modified type, and every virus of the modified type gave only progeny of the original T2 type when it reproduced in dysentery bacilli. What we had, in short, was a nonhereditary modification of the virus imposed by the host bacterium itself.

Within a few months workers in many laboratories found cases of host-induced modifications in all sorts of bacteriophages besides T2. There was one important difference, however. The modification of T2 is "nonadaptive;" that is, the modified virus cannot grow in the host that changed it. In most of the other cases the changes are adaptive: the changed virus can grow in the host that modified it but becomes unable to grow in a second host, and when occasionally a particle manages to overcome the restriction against growing in the second host, it immediately gives rise to fully adapted particles. Return to the first host erases the adaptation completely. The virus has no "memory" of any host but the very last. Each modification eliminates all the previous ones.

The discovery of the ability of bacteria to alter their parasites raised a number of questions. First of all, what property of a bacterium gives it this power? Clearly the answer lies in the genetic make-up of the bacterium. A single mutation in the common coli bacterium, for instance, transforms it into the mutant variety that modifies the T2 virus. A most remarkable thing is that viruses themselves sometimes bestow the virus-modifying property on bacteria. There is a latent form of virus known as "provirus," or "prophage" [see "The Life Cycle of a Virus," by André Lwoff; SCIENTIFIC AMERICAN, March, 1954]. The prophage, apparently incorporated in the chromosomes of the host bacterium and multiplying with them, occasionally turns into full-fledged virus and destroys the bacterium. Some prophages control the production of substances by their hosts (*e.g.*, diphtheria toxin) or have other important effects on them. Now two British bacteriologists, E. S. Anderson and A. Felix, have discovered that a prophage can cause certain typhoid bacilli to produce modifications in viruses completely unrelated to the prophage.

In our laboratory Seymour Lederberg has discovered recently that a single virus particle can possess two distinct host-induced modifications. The virus is first modified so that it can grow in a host in which it could not grow before. A second modification enables it to grow in a host containing a certain prophage. Both adaptations are reversible: they can be removed by letting the modified virus reproduce a new generation in an appropriate normal host.

These findings prove that a bacterium's modifying influence on a virus can be traced to specific portions of the host's hereditary material. Indeed, the prophage-controlled properties of bacteria may become extremely useful in the study of latent viruses and of gene action in general.

Exactly what are the changes that occur in a modified virus? We still do not know, but we can guess where to look for the differences between the original and the altered virus. The critical stage in the life cycle of a bacterial virus comes just after its hereditary material, the nucleic acid DNA, invades the bacterium. There is a good deal of circumstantial evidence that the injected virus material ordinarily establishes some contact with the nucleus of the host cell. There it takes one of two courses: it may become integrated with the host nucleus as prophage or it may begin at once to reproduce as virus. Now when a virus is modified in such a way that it cannot grow in a certain host, the halt in its development comes at this early stage. The virus's reproductive material penetrates into the host, but somehow it fails to make the proper adjustments for reproduction. It neither reproduces nor becomes prophage. The guess is that this failure is due to a change in the virus's nucleic acid which prevents it from establishing fruitful contact with the nuclear material of the host. One piece of evidence which may support this concept is that some modified viruses can be made to grow in unreceptive hosts by pretreating the host cells with ultraviolet light, which acts rather specifically on their nuclear apparatus and may facilitate successful contact.

Is it possible that modifications like those in bacterial viruses may occur in the viruses responsible for human diseases? We have no way of knowing so far; indeed, there is no evidence that the multiplication of viruses in animal cells is at all like the reproduction of bacterial viruses. Yet the Australian virologist H. J. F. Cairns has observed a suggestive parallel. When influenza virus grown in a chicken egg is transferred to the brain of a mouse, it multiplies only in the first batch of cells that it meets and no further. Cairns suggests that the brain cells may modify the virus in such a way that it ceases to be able to grow in such cells, though the modified virus can still grow in eggs—just as the modified T2 virus becomes unable to grow in the coli cells that produced it but can multiply in the dysentery bacillus.

This gives rise to some interesting speculations. If animal cells can modify viruses, they might well control the spread of viruses in animal tissues. Some viruses that have multiplied in certain organs can be stopped by others. We may even speculate about the possibility that there are viruses which transform normal cells into tumor cells and then are so modified themselves in the latter that they cannot reproduce further.

A new view of the nature of viruses is emerging. They used to be thought of solely as foreign intruders-strangers to the cells they invade and parasitize. But recent findings, including the discovery of host-induced modifications of viruses, emphasize more and more the Similarity of viruses to hereditary units such as genes. Indeed, some viruses are being considered as bits of heredity in search of a chromosome.

## ABOUT THE AUTHOR

■ **Salvador E. Luria** (1912-1991) was awarded the Nobel Prize in Physiology of Medicine in 1969 along with Max Delbrück and Alfred D. Hershey for their discoveries concerning the replication mechanism and the genetic structure of viruses.

# Skin Transplants

*When skin from one man is
grafted onto someone other than his
identical twin, it soon drops off. The chemical
mechanism which interferes with the graft
is illuminated by animal experiments.*

## P. B. Medawar

Skin-grafting was introduced into medical practice by Jacques Louis Reverdin, a surgeon in Paris, about 90 years ago. In principle it is quite a simple operation. The skin has two layers: an outer epidermis, consisting of tiers of cells which are constantly replaced from the inside outward, and an inner dermis, or corium, consisting mainly of a latticework of tough connective-tissue fibers, to which skin owes its great strength. The portion sliced off for grafting is made up of the epidermis and the upper part of the dermis. Its transplantation amounts to little more than laying it in place over the area to be repaired and holding it there under light but firm pressure until the dermis becomes knitted to the graft bed below.

The early plastic surgeons supposed that skin could be grafted from one person to another; some of them—victims of heaven-knows-what enormities of self-deception—convinced themselves that even the skin of frogs and rabbits could be transplanted to man. It was not until 1911 that Erich Lexer, in a masterly address before a conference of German surgeons, showed conclusively that skin grafts exchanged between different persons, even between parents and their children, were invariably unsuccessful. The truth of what Lexer said was slowly and grudgingly conceded. Nowadays everyone agrees that skin transplanted from one individual to another will not survive permanently. After a week or so the transplanted skin becomes puffy and inflamed, and soon the graft is sloughed off or drops away. Ordinarily the only kind of graft that will work is an autograft—that is, a transplant of an individual's own skin from one part of his body to another.

One exception to this rule is that skin can be exchanged between identical twins. This test has provided crucial evidence in cases of disputed or uncertain parentage, as two stories will illustrate. The first story, which has the makings of an operatic libretto, is about three six-year-old boys called Victor, Pierre and Eric. Victor and Pierre were supposedly twin brothers, and would still be so regarded if their father (as he imagined himself to be) had not had his attention called to a third boy, Eric, who was the very image of his Victor. Inquiry showed that Eric had been born in the same maternity clinic and on the same night as Victor and Pierre. It seemed likely that Eric was Victor's real twin and that Pierre had been substituted for Eric by mistake. A very careful physical comparison (including a study of finger prints, eardrum patterns and X-rays of the hands) made it virtually certain that Eric and Victor were identical twins. However, the mother to whom Eric had been allotted did not take kindly to the view that the boy whom she had brought up to the age of six was not in fact her son. Blood-group tests failed to exclude the possibility that Eric might be her son (though they did prove that Pierre could not be Victor's mother's son). It was agreed that a skin-grafting test would be decisive. A surgeon, Sir Archibald McIndoe, transplanted small squares of skin between Eric and Victor and between Victor and Pierre. The grafts exchanged between Pierre and Victor were sloughed off. But Eric and Victor accepted the grafts from each other, a result which proved that they must be identical twins. So Eric and Pierre were restored to their rightful and now satisfied parents, and the story ends happily (which makes it unsuitable for an operatic libretto after all).

The second story is about a mother and daughter and a question of "virgin birth." Members of the staff at University College in London often give lunch-hour lectures which may be attended by the press and the public, and in the course of one such lecture a geneticist gave certain reasons for supposing that parthenogenesis (development of the egg without fertilization) occurred in guppies and might not inconceivably occur in man. (Later evidence indicated that the guppy births were probably a case of self-fertilization.) A section of the press, no doubt animated by a sense of public duty, instituted a campaign to find an authentic example of virgin birth in human beings. In response to an appeal 19 mothers presented themselves with daughters—daughters they must be, for genetic reasons—of allegedly parthenogenetic birth. Eleven who had not quite grasped the import of the idea of virgin birth were at once eliminated; seven more were disqualified by differences between the mother's and daughter's blood types. But there remained one mother whose daughter qualified on blood grouping and certain other grounds. To clear the matter up, skin grafts were transplanted from the mother to the daughter and vice versa. Both grafts broke down and were sloughed away in a matter of weeks. The failure of the graft transplanted from the daughter to the mother proved that the child must have had a father.

The reaction which causes one individual to reject a graft from another is not a peculiarity of human beings. With the mysterious exception of the hamster, every species of vertebrate so far tested has exhibited this reaction against homo-grafts (*i.e.*, transplants between different individuals of the same species). Nor is the reaction confined to skin, though no other tissue shows it so clearly. W. J. Dempster of the Postgraduate Medical School of London has shown that it applies to a graft of a whole kidney, and it has also been shown to apply to the heart, the lung and even to grafts of tumor tissue. Some parts of the body will accept homografts—for example, the cornea of the eye and the brain—but in these cases special factors are at work.

Plainly the reaction against a graft is an immunological one; *i.e.*, a reaction of the same general kind as that provoked in the body by foreign proteins, foreign red blood cells, or bacteria. This is easily demonstrated by experiments. After a mouse has received and rejected a transplant from another mouse, it will destroy a second graft from the same donor more than twice as rapidly, and in a way which shows that it has been immunologically forearmed. This heightened sensitivity is conferred upon a mouse even when it merely receives an injection of lymph-node cells from a mouse that has rejected a graft.

In most immunological reactions the body employs antibodies as the destroying agent—*e.g.*, in attacking foreign proteins, germs and so on. Antibodies are formed in response to a homograft, but there are reasons to doubt that these are normally the instruments of the reaction against such a graft. Paradoxically enough, a high concentration of circulating antibodies seems if anything to weaken the reaction: it allows the graft to enjoy a certain extra lease of life.

The actual agents of attack on the graft seem to be not antibodies but cells produced by the lymph glands. Some skillfully designed experiments by C. H. Algire, J. M. Weaver and R. T. Prehn at the National Cancer Institute in Bethesda certainly point in that direction. In one experiment they enclosed a homo-graft in a porous capsule before planting it in a mouse which had been sensitized by an earlier homograft from the same donor. When the pores of the capsule were large enough to let cells through, the mouse destroyed the graft. But when the experimenters used membranes with pores so fine that they kept out cells and let through only fluid, the graft survived.

The hypothesis that the action against a graft is carried out by cells explains why grafts in the cornea are mercifully exempted from attack. The cornea has no blood vessels; consequently blood-borne cells cannot reach the graft. In the brain, on the other hand, the con verse of this situation obtains: the brain lacks a lymphatic drainage system, so that any antigens released by a graft there may not be able to travel to centers where they can stir up an immunological response. This probably explains why homografts can often be transplanted successfully into the brain.

For some years past at University College in London R. E. Billingham, L. Brent and I have been studying the cause of the reaction against homografts and steps that can be taken to prevent it. Following up a clue provided by the work of R. D. Owen at the California Institute of Technology, we discovered that the power to react against homo-grafts could be prevented from developing if we injected an animal at a very early age with cells from the donor strain—most conveniently cells of the spleen. In adult mice the injection of such cells increases the mouse's resistance to a graft from the donor. But if the spleen cells are injected in a mouse in the fetal stage or very shortly after its birth, the opposite happens: the mouse becomes tolerant of grafts from the strain that provided the spleen cells, though it remains intolerant of homografts from mice of other strains. A tolerant mouse can be recalled to a sense of the fitness of things by injecting it with lymph-node cells from a normal mouse of its own strain. Its tolerance then slowly disappears. The operation seems to equip the tolerant mouse with cells which are competent to recognize and react against foreign substances issuing from the previously tolerated homograft.

This experiment, among others, shows that the tolerance of a homograft is due to absence of specific reactivity in the host, rather than to any change in the properties of the grafted tissue. The antigens are present in the graft, but the animal cannot react to them. The phenomenon of tolerance of antigens cannot yet be explained by any chemical theory of the immunological reaction. No future theory will be acceptable unless it can take tolerance in its stride.

What are the tissue antigens that cause an animal to reject a homograft? When we began our work, it was known only that they are very numerous and that they are under the most exact genetic control. While the antigens could not be identified chemically, they could be separated by genetic methods, i.e., by a combination of breeding and grafting tests. E. J. Eichwald and C. R. Silmser at the Deaconess Hospital in Montana have just made the remarkable discovery that, within certain inbred strains of mice, a female will accept a skin homograft from a female, a male from a male and a male from a female, but a female will not take a graft from a male. The discovery raises the possibility that the Y chromosome of male mice, hitherto thought to be concerned only with sexual differentiation, may actually control the formation of an antigen.

The study of the nature of the antigens that cause the homograft reaction is difficult and laborious. Their action cannot be investigated in the test tube but only by effects on living animals. Moreover, the antigens are highly unstable. Cells which have been frozen and thawed or dried in the frozen state or heated to about 120 degrees Fahrenheit are no longer capable of eliciting a homograft reaction. Fortunately we have discovered ways of disintegrating cells without destroying their antigenic power. For example, with judicious use of ultrasonic radiation we have broken down cells into nuclear and cytoplasmic fractions and found that the antigenic power lies only in the nuclear material. The antigenic substances in the nucleus are not soluble in water of the same salinity as the body fluid (*i.e.*, about 1 per cent sodium chloride). But they can be coaxed into solution in distilled water. After we learned the trick of dissolving them, we found by centrifuging tests that the active substances either were very large molecules or very small particles or were firmly attached to a large molecule or particle. The active matter can be precipitated from the water solution by adding a very small amount of magnesium chloride to the solution. It can be partly redissolved by raising the concentration of magnesium chloride and can then be precipitated again simply by adding water to dilute the solution. I mention these reactions because they have the crispness and clarity that one usually associates with schoolroom chemistry, and because of the commendable frugality of the reagents which we use—water and a number of simple salts.

Our experiments suggest, but do not yet prove, that the antigens responsible for the homograft reaction are compounds of desoxyribonucleic acid and protein—that is, chromosomal matter. Tests on the breakdown of the antigenic substances by specific enzymes support this interpretation. Furthermore, the idea that these antigens are chromosomal matter fits well with evidence that the antigens are present in all the tissues of the body, in embryos and (so it has been said) even in sperm.

Desoxyribonucleic acid itself does not act as an antigen. It is active only in combination with protein. Perhaps the protein part, like the protein coat of a bacterial virus, simply helps the nucleic acid to get into the cells from which it elicits a reaction—in this case, the lymphoid cells of the host. If that is so, the nucleic acid may be the part of the antigen that is specifically responsible for its power to sensitize an animal against the cells of a foreign graft. We have not yet proved this, but the evidence that desoxyribonucleic acid is the chemical embodiment of heredity makes it a plausible guess.

If our interpretation turns out to be true, the study of the antigens that cause the homograft reaction could be of decisive importance in working out the chemical structure of chromosomes, for it would provide a test to determine whether a chromosomal extract has been damaged by the process of extraction. At present no one can be sure, for only in microorganisms do we have biological tests that can guarantee that the nucleic acids are in working order. Our hypothesis has many other deeper and more exciting implications also, but there will be time enough to consider these when we have assured ourselves that the hypothesis is correct.

## ABOUT THE AUTHOR

**P. B. Medawar** (1915-1987) was awarded the Nobel Prize in Physiology or Medicine in 1960 along with Sir Frank Macfarlane Burnet for their discovery of acquired immunological tolerance.

# The Structure of the Influenza Virus

*A sequel to "The Influenza Virus" published in the April, 1953, issue of this magazine. Since that time the behavior of the virus has been increasingly related to its physical and chemical nature.*

## Sir Macfarlane Burnet

Nearly 40 years ago the influenza virus took on an exceptional virulence and within less than two years caused, directly and indirectly, at least 50 million deaths. In 1918 and the decade following, influenza was the great unsolved mystery in the field of infectious diseases: the isolation and taming of the virus was the greatest prize that any bacteriologist could strive for. In the next three decades much of the mystery was removed from the disease. The influenza virus was isolated, first in swine by Paul A. Lewis and Richard E. Shope at the Rockefeller Institute for Medical Research in 1931 and then in the form infecting man by Wilson Smith, Christopher H. Andrewes and Patrick Laidlaw in England in 1933. Various methods of studying the virus in the laboratory were developed. It was found that influenza virus could be grown readily in fertile chicken eggs, and ways of measuring accurately the amount of virus grown were discovered.

Thus we have today a most convenient system for studying the character and activity of the influenza virus. Not the least convenient feature of the system is that after the virus has been cultivated in chick embryos it seems to lose all its virulence for human beings. It is almost a joke that nobody in an influenza virus laboratory ever catches influenza from his cultures. Yet in all essentials this domesticated, harmless virus is still influenza virus. Anything that can be learned of how it multiplies in the cells of the chick embryo may one day be relevant to matters of life and death.

Basically the question we are now seeking to answer is: What is the influenza virus and how does it become what it is? At one moment we have a single virus particle and a normal living cell. Some hours later we find many virus particles plus a damaged and dying cell. The problem is to understand how this takes place. In investigating the matter we have three different entities to consider: the virus particle itself, the host cell and the infected cell, which is more than a simple association of the two. The virus particle has a relatively simple structure which we may hope to understand fairly well. The normal living cell is something of which our knowledge is both enormously extensive and utterly incomplete. The infected cell presents us with a far more complex problem. It is perhaps characteristic of the growing edge of biology that when a new phenomenon like virus multiplication comes to be studied, almost all the knowledge of cellular chemistry and function gained from other types of study turns out to be irrelevant. Any attempt to picture what is happening in the infected cell must necessarily therefore be provisional and oversimplified. At best it can provide only a framework which makes it easier to grasp relations and sequences and to devise approaches to more adequate knowledge.

To study the production of virus experimentally, the first need is a means of measuring the amount produced under given conditions. Three methods of measurement have been developed. The most direct is an assay of the infectivity of the fluid containing the virus. In the fluid taken from a chick embryo two or three days after it has been infected, there is a very large amount of virus. If we dilute this fluid and inoculate fresh chick embryos, they will show the characteristic results of infection with all dilutions up to perhaps one part in 100 million. At a certain dilution exactly half the embryos will show infection. This method, suitably elaborated, serves to establish a unit of infectivity which gives a measure of the amount of virus present in the original fluid.

The second method of estimating the amount of virus is to count the virus particles made visible with the electron microscope. In suitable preparations an infected fluid can be seen to contain characteristic particles whose number is closely correlated with the degree of infectivity of the fluid.

The third method uses the hemagglutinin reaction (clumping of red blood cells) developed in 1940 by George Hirst at the Rockefeller Institute. Influenza virus particles have a special surface quality which makes them stick rather firmly to the surface of the red blood cell. If virus-containing fluid is mixed with a suspension of red cells (say from chicken blood), the virus particles stick to any cell with which they collide and can act as bridges holding two red cells together. Provided there are at least as many virus particles as red cells in the mixture, the cells will be tied together into clumps. These clumps settle to the bottom of the tube much more rapidly than normal cells. The hemagglutinin reaction is a very convenient tool for measuring the amount of virus in a given fluid, as well as for many other uses in biology.

What, then, do we learn of the influenza virus once we have grown it in sufficient amount for analysis? Chemically the virus is found to be built up of materials common to most living cells. Around 30 to 40 per cent of it consists of the same lipids (fatty substances) as are found in vertebrate cells. It contains proteins and carbohydrates which are indistinguishable by gross chemical methods from similar material in the chick embryo cells in which the virus was grown. The only component in which the virus differs chemically from the host cell is its nucleic acid. In the virus the sole nucleic acid present is ribonucleic acid (RNA), and Gordon L.

Ada of our laboratory in Melbourne has shown that the RNA of the virus is chemically distinct from the RNA of the host.

If the virus particle is almost indistinguishable from a tiny blob of protoplasm chemically, it is nonetheless highly individual functionally. Aside from its capacity to infect a cell, it has other distinctive activities. It sticks to red blood cells because it carries molecules which fit appropriately to certain complex compounds of carbohydrate and protein (mucoprotein) which are present on the surface of all vertebrate cells. Virus particles in time can liberate themselves from their attachment to cells by means of an enzyme which breaks off the portion of the mucoprotein to which they are held. Furthermore, they react in specific ways to antiserum. Serum from an animal or man convalescent from influenza will inactivate one sort of influenza virus, but is ineffective against any other type.

It is easy to see that by a series of only moderately complex manipulations with red cells, mucoprotein solutions and antiserums, one can learn a great deal about individual differences between viruses and by implication about the protein molecules on their surfaces which are responsible for the various reactions. We can summarize what such experiments have shown by saying that each type of virus has its own structure, consisting of protein molecules arranged probably in a more or less symmetrical pattern over its surface with lipid and mucoprotein molecules in between. The pattern of these protein molecules is determined genetically, and we believe it must be carried by the virus's RNA.

In the electron microscope we can see two differently shaped kinds of virus particles. The most common kind are spheres, about a tenth of a micron in diameter. But often we see particles in the shape of filaments, sometimes as long as 20 microns. The spheres are known to be infective. The filaments, however, appear to have the power to infect only when they contain a "knob," usually at one end.

Recently Councilman Morgan, Harry H. M. Rose and Dan H. Moore of the College of Physicians and Surgeons have produced some magnificent pictures of virus multiplication, and these tell us some important things. First of all, the virus particles that emerge from an infected cell appear only at a free surface of the cell, where it is exposed to fluid and is not in contact with another cell. At the peak of multiplication, buds of virus units can be seen along the free surface of the cell, and immediately below the surface there are little opaque concentrations, which are clearly fated to become virus particles once they pass through the surface layer.

The pictures suggest too that the virus particle actually derives its own outermost coat from the membrane of the cell. We get the impression that a filament is produced when the cell surface membrane fails to close neatly around an emerging virus particle, so that the virus core, instead of becoming a spherical virus, goes on pulling out an indefinitely long cylinder of material.

The filaments are far too thin to be seen by standard light microscopes, but when they are examined by indirect dark-ground microscopy at a magnification of about 500 times, they are easily visible as flexible lines of light. We can count them and determine their length and formations. This allows us to note changes in these features and thereby carry out various experiments which would be difficult to manage if we had to use the complicated manipulations needed for electron microscopy.

Only one or two such types of experiment need to be mentioned. We have found that every physical or chemical agent which causes red blood cells to release their hemoglobin will also produce visible damage to filaments under the same conditions. Sometimes the filaments disappear altogether; more often their number decreases and those remaining appear irregularly angulated and abnormally rigid. The effective agents include heat, distilled water, ether, chloroform, many detergents and such substances as saponin and cobra venom. The suggestion is strong that the surface of the filament is basically very like the surface of an animal cell.

In nearly every respect the surface of a filament has the same qualities as that of a spherical virus particle. Even a little piece of broken filament can attach itself to a red cell and later release itself by enzyme action. Mix an antiserum against influenza with a suspension of filaments and the filaments clump into fluffy balls, showing that they have the same characteristic protein antigen on their surface as the normal influenza virus does. We are almost compelled, therefore, to regard the surface of the filament as a mosaic of components, some from the host cell, others whose structure is determined by the virus. There is in fact much to suggest that in the surface of a filament, the protein is virus protein, but the lipids and carbohydrates are unchanged components derived from the cell membrane.

There is one final point about filaments that is vital to their understanding. The production of filaments is a hereditary trait: some viruses produce a high proportion of them, others a low proportion. We have completed experiments which show that the power to produce filaments can be lost by mutation rather readily, other recognizable characters remaining the same.

In the course of genetic experiments, we have been able to produce "hybrid" viruses. This is accomplished by infecting the same cell with two distinct viruses. We can take two viruses which differ in their antigenic pattern so that each is inactivated by a different serum but neither serum is effective against the opposite type. We then arrange things so that in a group of cells each is infected with at least one virus particle of each type and we collect the new virus liberated from these cells seven or 10 hours later. If each new-formed virus particle has inherited its traits from only one parent, we should find that one antiserum inactivates about half of the progeny and the other serum also about half. When the test is made, however, we find that either serum inactivates about 90 per cent of the new virus. The only possible interpretation is that the great majority of the new virus particles partake of the surface properties of both types of parental viruses. This suggests very strongly that both parents have contributed to what may be called a pool of virus components from which the new virus particles are fabricated.

The process by which a new generation of virus particles is produced in the infected cell will never be fully understood; the complexity of the cell processes concerned will see to that. But at a fairly superficial level a useful picture of the process is already emerging. Much more has contributed to this picture than we have been able to mention in this article.

When an infectious virus particle makes contact with a susceptible cell, the specially patterned proteins on its surface become attached to mucoprotein of the cell. The virus then sinks into the cell, and there it unfolds itself in a way which destroys its individuality as an identifiable virus particle but allows the vital carrier of that individuality—its RNA—to make contact with the synthetic mechanisms of the cell. At this stage we have what is really the intrusion of one genetic mechanism into a territory shaped and controlled by another. It is not too fantastic an analogy to compare the virus to a sperm entering the ovum. In each case the intrusion results in the cell taking on a new character which depends as much on the intruder as on the occupier.

We have only the most general picture of what ensues in the infected cell, but we do know that the total amount of RNA in the cell increases, and that recognizable virus protein begins to appear at an early stage. It seems that there develops in the cell a "replicating pool" of virus components, especially nucleic acid and protein. We can picture a swirl of activity by which the chemical building blocks needed for synthesis of the giant molecules are brought into position by the cell's normal synthetic mechanisms, but because of the intrusion of the virus, the newly formed protein and nucleic acid are built to the virus pattern instead of the host pattern.

Out of this turmoil, appropriate groups of virus molecules come together to produce what can without too much violence to the meaning of the word be called "nuclei" of viruses. Simultaneously viral protein molecules must move into the cell membrane to produce a new fabric, as it were, with which to coat the virus nuclei as they emerge from the cell. It seems likely that there is some spatial regularity in the distribution of virus molecules amongst the lipid and carbohydrate constituents of the cell surface. Perhaps there are two types of spatial symmetry—one which is best satisfied by a spherical distribution and another which determines the indefinite extrusion of a cylindrical filament.

So we can catch a glimpse in broad outline of the process by which the new generation of virus particles emerges from the cell. If we look at the process from an even broader point of view, we can perhaps summarize it as a continuing alternation between two modes of life. A virus is not an individual organism in the ordinary sense of the term but something which could almost be called a stream of biological pattern. The pattern is carried from cell to cell by the relatively inert virus particles, but it takes on a new borrowed life from its host at each infection.

## ABOUT THE AUTHOR

**Sir Frank Macfarlane Burnet** (1899-1985) was awarded the Nobel Prize in Physiology or Medicine in 1960 along with Peter B. Medawar for their discovery of acquired immunological tolerance.

# The Mechanism of Immunity

*How does an animal make an antibody
that neutralizes a single foreign substance,
or antigen? The evidence favors the theory that
cells able to make the antibody are "selected"
by the antigen and then multiply.*

## Sir Macfarlane Burnet

**D**eliberate defense against infectious disease started in the late 18th century with Edward Jenner's discovery of the principle of immunity, so triumphantly demonstrated by the success of Jenner's vaccine against smallpox. Today the technique of immunization provides protection against all the significant diseases that have not been eliminated by public-health measures or that do not yield readily to chemotherapy. Much public-health work remains to be done, particularly in the underdeveloped areas of the tropics, but without important exception man can now control all the infectious diseases that seriously threaten human life.

Although the practical problems of immunization have been solved, immunology remains an important branch of medicine. The immunologist of today, however, is not so much interested in finding out how to immunize people more effectively against diphtheria or poliomyelitis as he is concerned with understanding what happens when people become immune. He asks more sophisticated questions than in the past. For example: Why can a surgeon successfully graft skin or other tissue from one part of the body to another but not from one individual to another, except in the case of grafts between identical twins? How is it that occasionally a pair of fraternal (nonidentical) twins share two blood groups and accept skin grafts from each other? How can an individual who had suffered a single attack of a virus disease 20, 30 or even 60 years ago continue to produce antibody against the virus? And why are there "autoimmune" diseases, such as rheumatoid arthritis, acquired hemolytic anemia and Hashimoto's disease of the thyroid, in which an abnormal immune reaction is directed against the body's own cells and tissues? Any modern formulation of immunological theory must supply at least provisional answers to these and other equally complex questions.

But immunology is not simply a branch of medicine. It is a discipline in its own right, potentially able to make a rich contribution to the understanding of the central problems of biology, notably the nature of genetic information and the mechanism of protein synthesis. Both of these problems are intimately tied to any theory of immunity.

In its modern form orthodox immunological theory holds that the central feature of immunity is the production of antibody by a specialized group of tissue cells known as plasma cells. Antibody is a globular protein of blood plasma which can be identified by its physical behavior as a "gamma globulin." Each antibody has a highly specific affinity with the particular antigen which stimulates its production. An antigen may be part of a virus, bacterium or foreign tissue cell, or a fragment of some such structure, a protein or a polysaccharide (a large molecule made of many simple sugar units). Antibody protects the organism against a foreign substance by combining with it and thereby rendering it inactive.

Antigen and antibody are both large in the chemical sense, that is, the molecules of both consist of a great many atoms. Antibody globulin has a molecular weight of about 160,000 (10,000 times the weight of the oxygen atom). Typical antigens are of the same order of size. The sites of chemical activity which bring antibody and antigen together into combination, however, represent relatively small portions of these complex molecules. A single site may be thought of as equivalent to the region occupied by three to five of the several hundred amino-acid units in an average protein (a protein being composed of combinations of any of 20-odd different amino acids), or an equally small number of the monosaccharide units in a polysaccharide. These small regions of active union are called antigenic determinants on the antigen and specific patches on the antibody. According to the classical theory, the two combine because the geometrical configuration of the specific patch is complementary to the pattern of the antigenic determinant. They fit each other just as a particular key matches its lock. In this scheme, which bears the strong imprint of such figures as Paul Ehrlich, Karl Landsteiner and Linus Pauling, the specific patch on the antibody acquires its pattern by being synthesized in contact with the antigenic determinant. The antigen itself is presumably taken into the cell and comes into action after the amino-acid units of the globulin molecule have been assembled by the cell's machinery of synthesis and are in process of being folded into globular form. At the folding stage the globulin is brought into contact with the antigen and is molded into the required complementary pattern.

This is the simplest form of what Joshua Lederberg of Stanford University has called the "instructive" theory of antibody formation: the antigenic determinant itself supplies the information from which each highly specific antibody is constructed. The instructive theory does not, however, account satisfactorily for several significant processes associated with immunity, such as the persistence of immunity and the origin of the autoimmune diseases. A fundamentally different view has accordingly been advanced by the proponents of the so-called selection theory.

This theory holds that antibody molecules are made in essentially the same way that other proteins are synthesized, that is, according to genetic instructions contained in the nucleus of the manufacturing cell. At no time does information enter the cell from outside. Instead, for each one of the thousands of possible foreign antigens, the body already contains a cell or group of cells genetically capable of synthesizing the appropriate antibody. Each of these cells or groups of cells "knows" how to make the specific antibody even if the complementary antigen never enters the body. The function of the antigen is simply to select and stimulate the proliferation of the appropriate group of cells, thus increasing production of the required antibody.

The idea of selection has been central to biology ever since the publication of the *Origin of Species:* The environment selects among organisms for the differential survival attributes or potentialities which are conferred upon them by genetic processes. The sun does not breed maggots in a dead dog unless the fertilized fly deposits the necessary genetic information in the carrion. No one now seriously argues that evolution produced the whale and the giraffe by the Lamarckian formula according to which function in the environment molds form—first physically and then inheritably—in the right direction. Recently, however, some investigators have held that bacteria show a wide capacity to produce "adaptive" enzymes on demand. It was indeed observed that bacterial cultures can start producing new enzymes when presented with unusual substances in their nutrient. But it soon became clear that adaptive-enzyme formation is a much subtler phenomenon. Current interpretation tends toward the view that a bacterium can produce a given enzyme only if the necessary information is incorporated in its genetic mechanism; the experimental change in the environment allows the emergence into activity of what was formerly only a latent capacity.

It is likely that views of antibody formation will change in the same direction, toward a wider acceptance of selection theories. Certainly this approach leads more directly to the central process in immunology, which I defined a long time ago as the differentiation between the self and the not-self. The body does not normally produce antibodies against its own tissues, although it is at least potentially capable of producing antibodies against any protein or any other substance of appropriate molecular character that is not present in the body. The implications of this fact are the most important reasons for favoring a selection theory of immunity.

Most proteins are antigenic to an organism that has not been concerned in producing them. At the present time only one protein is known well enough to permit a comparison of its chemical structure with its immunological activity. This is insulin, one of the smallest proteins; the full sequence of amino-acid units has been worked out for the insulins of several animal species. Of course insulin is not antigenic in the animal that produces it, and it also happens that it is a rather mild antigen. Most diabetics can receive beef or sheep insulin for years without any trouble. Some, however, become resistant to insulin because they are making antibodies against it. This difficulty can usually be circumvented by using pig insulin. The difference between beef insulin and pig insulin is known. Of the 51 amino-acid units in the insulin molecule, 48 have the same arrangement in the insulins of different species; the sequence of units in one segment of three units varies. If an insulin is antigenic for a mammal, it is because this sequence differs from the corresponding sequence of the animal's own insulin.

In these days when genetics has come so close to biochemistry it is worth pointing out that an antigen, like a gene, is a purely relative concept. A gene is an entity devised to explain an observable hereditary difference between two interbreeding stocks or individuals. Long stretches of a chromosome must remain genetically silent if there are no regions of observable difference between available stocks. An antigen, or more strictly an antigenic determinant, is also an expression of difference. It contains certain patterns which differ from any pattern present in the animal in which it is tested for antigenicity. In one kind of animal one part of a foreign protein molecule may be antigenic; in another species an entirely different segment of the same molecule may stimulate the antibody reaction.

Even though insulin is a poor antigen, it still presents rather clearly the central question: How does the insulin-resistant diabetic "recognize" the tiny difference between beef insulin and his own insulin and so make antibody against the former? Basically this is a problem of information. How does the body acquire or generate the information needed to distinguish foreign chemical configurations from its own?

The most important clue is provided by experimental manipulations that trick an organism into accepting as its own a substance or a cell that genetically speaking has no right to be there. Probably the most impressive example comes from the rare experiments of nature by which genetically dissimilar human twins share a common placental circulation in their mother's uterus. This will ensure that each twin receives a variety of cells from the other, including cells that can settle down in the bone marrow and multiply to produce the red blood cells. Three pairs of such twins have been recognized in adult life. When their blood was typed prior to their acting as blood donors, they were found to have two blood groups: their own and that which was genetically appropriate to their twin. Such fraternal twins have a second striking difference from an ordinary pair of dissimilar twins. Fraternal twins who have developed in the usual fashion from two separate placentas will not accept skin transplants from each other. An immune reaction kills the grafts. But fraternal twins with double blood groups (at least the only pair of twins so far tested) have been found to accept cross skin-grafting as happily as if they were genetically identical twins. This indicates that self-recognition in the antibody-producing system is not due simply to hereditary traits. Rather, self-recognition seems to develop as a secondary process sometime during embryonic life.

Much work has been done in recent years on the experimental demonstration of immunological tolerance, most often in mice and rats. Laboratories now possess many lines of mice so inbred and so similar genetically that each individual will accept grafts of skin or other tissue from any other member of its strain. Two very illuminating experiments can be carried out with two suitable strains: A and B. In both experiments an emulsion of living cells from a mouse of strain B is inoculated into a vein on the face of a newborn mouse of strain A. This requires steady hands and a good eye, but it is done routinely.

In the first experiment cells from the spleen and kidney of an embryo of mouse B are inoculated into a newborn mouse of strain A. The mouse develops normally. If a piece of B skin is grafted to the mouse when it is sufficiently grown, the graft "takes" and persists in a healthy condition. If the A mice are white and the B mice are black, the A mouse presents the unprecedented anomaly of a patch of healthy black hair.

In the second experiment another mouseling of strain A is inoculated with cells from the spleen of an adult B mouse, not an embryo. The result is disastrous. Depending upon the number of cells and the particular pair of mouse strains being used, the mouseling either dies within two or three weeks or develops slowly into an undersized, scruffy-looking individual suffering from what has been called runt disease.

A slightly oversimplified explanation is that in the first experiment host A becomes tolerant of the B cells implanted in its tissues just after birth. As a result A subsequently tolerates a graft of B skin. But it is important to note that the cells that are implanted have qualities just as definite as those of the host. If an equilibrium is to be reached, the implanted B cells must become tolerant of their foreign host as well as vice versa. The embryonic B cells do become tolerant. But in the second experiment the adult B cells set up their own immune reaction against their host and produce runt disease or death.

A detailed consideration of many phenomena of the same general quality permits the formulation of the key question in the self and not-self problem: What is the process by which the body learns during development to differentiate its own substance from that of others? As Niels K. Jerne of the World Health Organization has put it, where or what is the dictionary that the body must consult to decide whether such and such a word (chemical configuration) is foreign or belongs to its own language? Along with Lederberg and other investigators, he believes that the dictionary lists only foreign words and that it has in it a list of all the foreign words without ever having heard them!

Such a dictionary can be pictured in several possible ways, but basically it must contain a large, though not infinite, number of patterns (words) which among them can offer a complementary specific antibody patch to correspond with every possible antigenic determinant. The proposal is not as outrageously unlikely as it sounds, because the number of antigenic determinants is not impossibly large. Both the antigenic determinants and the specific patches are small configurations. The number of different three-

and four-letter combinations for the 20-letter alphabet of amino-acid units in proteins is respectively 8,000 and 160,000; these are very few compared with the number of cells in even a mouse. David W. Talmage of the University of Colorado School of Medicine has estimated that only some 10,000 different patterns are needed.

It is not difficult to imagine how the body might create its foreign-word dictionary. The lymphocytes (one important group of the mobile white blood cells) are the most likely carriers of the words, or antibody patterns. In the early stages of embryonic life the ancestors of these cells are assumed to be highly mutable in this particular quality. Their genetic material would change spontaneously and in a random fashion, creating all the possible antibody patterns. Each mutated cell, through simple division, would become the ancestor of a small group of cells, called a clone, all identical and all carrying the pattern for one or at most a few specific antibodies. Since the mutation process would be random, antibody patterns against antigens within the body would arise. It is therefore necessary to postulate that such cells are destroyed by contact with their corresponding antigen. (It is well known that a high concentration of antigen in an adult will inhibit antibody formation.) Thus during an early phase of embryonic development, "forbidden" clones that match "self-antigens would be eliminated as they arose. Foreign antigens normally cannot reach an embryo, but when they do (as in the case of the nonidentical twins sharing one placenta), they come to be accepted as self. If no foreign antigens reach the embryo, it presumably retains all the foreign patterns.

Later in embryonic life the rate of mutations in immunological cells would decrease drastically to the mutation rate found everywhere in the body throughout life. (It has been estimated that as many as a million body cells undergo mutation each day.) Forbidden clones would continue to arise, though infrequently, and would normally be killed off, or at least inhibited, while still immature. Mature immunological cells, instead of being destroyed by the appropriate antigen, would be stimulated by it to proliferate, producing among their offspring a great many of the plasma cells that probably manufacture the actual antibody molecules to combine with and deactivate foreign antigens.

The theory is called the clonal-selection theory because the action of the antigen is simply to select for proliferation that particular clone of cells which can react with it. In the original form of the hypothesis each clone was believed to carry only one pattern, but two patterns per clone now seems to accord better with evidence from observations.

Many immunologists are highly sympathetic to the general idea of a clonal-selection theory, but are skeptical of the necessity of limiting the capacity of a given cell or clone to one, two or at most three patterns. They would prefer a substantial number, perhaps 10 to 20 related patterns per clone. Some even press the idea to its logical conclusion and assume that every cell which is a potential antibody producer carries its own complete foreign-word dictionary and can therefore recognize any antigenic determinant and through its descendants produce antibody against it.

The main virtue of the clonal-selection theory in which not more than three antibody patterns are carried by a single cell or clone is that it is in principle and in practice amenable to a variety of experimental tests. It can be proved wrong. So far no one has offered an experimental means of differentiating between an instructive theory and the theory that every immunological cell carries all possible antibody patterns. Furthermore, it is extremely difficult to picture how every one of these cells could contain all the information needed for the recognition of every foreign antigenic determinant.

In biology the only function of a generalization is to present clearly a statement in such a form that any interested worker can grasp the type of experimental or observational information that will be needed to disprove it or to compel its modification. No theory can ever be proved to be correct. The only major virtue of the clonal-selection hypothesis is that it draws attention to the essential role of cells, rather than of antigens, in antibody production and immunity. Hence it must stimulate attempts to define the potentialities of single cells and to analyze the population dynamics of the immunological cells in the body.

Several experimental approaches are possible. One is the study of the ability of single cells to produce antibody and of the number of types of antibody that a single cell can produce. It is now established that in an animal immunized with more than one antigen most cells produce only one type of antibody, but that an occasional cell can undoubtedly produce two. Another type of experiment is based on finding a situation in which a small proportion of cells with some special antibody pattern can be sorted out from a large population of immunological cells. At our laboratory in Melbourne we have recently been engaged in producing immunological "pocks" on the chorioallantois of the chick embryo, the membrane in the egg which supplies the chick with oxygen. We do this by inoculating the membrane with white blood cells from a mature chicken. The inoculation produces one focus, or pock, for roughly every 20,000 white cells. One focus, we believe, represents one cell; our provisional interpretation is that this ratio of 1 to 20,000 reflects the proportion of the white cells with the preformed patterns that correspond to and react with antigens in the chick embryo not present in the chicken that provided the white cells.

The clonal-selection theory could be decisively disproved if it were possible to grow antibody-producing cells in tissue culture and show that from a very small initial population of cells any desired type of antibody could be produced by stimulation from a variety of antigens. So far no one has produced such a demonstration.

To me the most gratifying feature of the clonal-selection theory has been the way in which it can fit all the pieces into a reasonably self-consistent pattern. The explanation of the important autoimmune diseases was particularly obscure when the instructive theories of immunity were the only ones available. The selection theory allows these phenomena to fall easily into place. It postulates that a forbidden clone, through mutation or otherwise, enjoys an abnormal protection from destruction or inhibition by its corresponding antigen. There are difficult problems to be faced in some of the more severe autoimmune diseases, but in one group the process is more readily understood because the antigens concerned are normally of very limited accessibility in the body. They arise in such well-"insulated" tissues as those of the nervous system or the interior of the thyroid gland; they do not normally circulate in quantity in the blood and hence fail to eliminate the complementary clone during the embryonic selection period.

Once the antibody-forming cells start to attack, they break down the cells and tissues containing the antigen, releasing more antigen. The antigen stimulates proliferation of the forbidden clone, which steps up the attack, and the vicious circle of autoimmune disease sets in.

At the other, theoretical, end of the conventional range of medicine is the central problem of biology—the way in which genetic information in the chromosomes of the cell nucleus is expressed in the specific geometric configuration of proteins such as enzymes. At this level, too, the idea of a preadapted pattern determined by the genetic material, which is the essence of the clonal-selection theory, seems to fit better with modern conceptions of protein synthesis than the rather crudely mechanical concept of the orthodox instructive theory.

## ABOUT THE AUTHOR

- **Sir Frank Macfarlane Burnet** (1899-1985) was awarded the Nobel Prize in Physiology or Medicine in 1960 along with Peter B. Medawar for their discovery of acquired immunological tolerance.

# Hemoglobin Structure and Respiratory Transport

*Hemoglobin carries oxygen from the lungs to the tissues and helps to transport carbon dioxide back to the lungs. It fulfills this dual role by clicking back and forth between two alternative structures.*

## M. F. Perutz

*Why grasse is greene, or why our blood is red.*
*Are mysteries which none have reach'd unto.*
*In this low forme, poore soule, what wilt thou doe?*
—John Donne, "Of the Progresse of the Soule"

W hen I was a student, I wanted to solve a great problem in biochemistry. One day I set out from Vienna, my home town, to find the Great Sage at Cambridge. He taught me that the riddle of life was hidden in the structure of proteins, and that X-ray crystallography was the only method capable of solving it. The Sage was John

Desmond Bernal, who had just discovered the rich X-ray-diffraction patterns given by crystalline proteins. We really did call him Sage, because he knew everything, and I became his disciple.

In 1937 I chose hemoglobin as the protein whose structure I wanted to solve, but the structure proved so much more complex than any solved before that it eluded me for more than 20 years. First fulfillment of the Sage's promise came in 1959, when Ann F. Cullis, Hilary Muirhead, Michael G. Rossmann, Tony C. T. North and I first unraveled the architecture of the hemoglobin molecule in outline [see "The Hemoglobin Molecule," by M. F. Perutz; SCIENTIFIC AMERICAN, November, 1964]. We felt like explorers who have discovered a new continent, but it was not the end of the voyage, because our much-admired model did not reveal its inner workings: it provided no hint about the molecular mechanism of respiratory transport. Why not? Well-intentioned colleagues were quick to suggest that our hard-won structure was merely an artifact of crystallization and might be quite different from the structure of hemoglobin in its living environment, which is the red blood cell.

Hemoglobin is the vital protein that conveys oxygen from the lungs to the tissues and facilitates the return of carbon dioxide from the tissues back to the lungs. These functions and their subtle interplay also make hemoglobin one of the most interesting proteins to study. Like all proteins, it is made of the small organic molecules called amino acids, strung together in a linear sequence called a polypeptide chain. The amino acids are of 20 different kinds and their sequence in the chain is genetically determined. A hemoglobin molecule is made up of four polypeptide chains, two alpha chains of 141 amino acid residues each and two beta chains of 146 residues each. The alpha and beta chains have different sequences of amino acids but fold up to form similar three-dimensional structures. Each chain harbors one heme which gives blood its red color. The heme consists of a ring of carbon, nitrogen and hydrogen atoms called porphyrin, with an atom of iron, like a jewel, at its center. A single polypeptide chain combined with a single heme is called a subunit of hemoglobin or a monomer of the molecule. In the complete molecule four subunits are closely joined, as in a three-dimensional jigsaw puzzle, to form a tetramer.

## Hemoglobin Function

In red muscle there is another protein, called myoglobin, similar in constitution and structure to a beta subunit of hemoglobin but made up of only one polypeptide chain and one heme. Myoglobin combines with the oxygen released by red cells, stores it and transports it to the subcellular organelles called mitochondria, where the oxygen generates chemical energy by the combustion of glucose to carbon dioxide and water. Myoglobin was the first protein whose three-dimensional structure was determined; the structure was solved by my colleague John C. Kendrew and his collaborators.

Myoglobin is the simpler of the two molecules. This protein, with its 2,500 atoms of carbon, nitrogen, oxygen, hydrogen and sulfur, exists for the sole purpose of allowing its single atom of iron to form a loose chemical bond with a molecule of oxygen ($O_2$). Why does nature go to so much trouble to accomplish what is apparently such a simple task? Like most compounds of iron, heme by itself combines with oxygen so firmly that the bond, once formed, is hard to break. This happens because an iron atom can exist in two states of valency: ferrous iron, carrying two positive charges, as in iron sulfate, which anemic people are told to eat, and ferric iron, carrying three positive charges, as in iron oxide, or rust. Normally, ferrous heme reacts with oxygen irreversibly to yield ferric heme, but when ferrous heme is embedded in the folds of the globin chain, it is protected so that its reaction with oxygen is reversible. The effect of the globin on the chemistry of the heme has been explained only recently with the discovery that the irreversible oxidation of heme proceeds by way of an intermediate compound in which an oxygen molecule forms a bridge between the iron atoms of two hemes. In myoglobin and hemoglobin the folds of the polypeptide chain prevent the formation of such a bridge by isolating each heme in a separate pocket. Moreover, in the protein the iron is linked to a nitrogen atom of the amino acid histidinem, which donates negative charge that enables the iron to form a loose bond with oxygen.

An oxygen-free solution of myoglobin or hemoglobin is purple like venous blood; when oxygen is bubbled through such a solution, it turns scarlet like arterial blood. If these proteins are to act as oxygen carriers, then hemoglobin must be capable of taking up oxygen in the lungs, where it is plentiful, and giving it up to myoglobin in the capillaries of muscle, where it is less plentiful; myoglobin in turn must pass the oxygen on to the mitochondria, where it is still scarcer.

A simple experiment shows that myoglobin and hemoglobin can accomplish this exchange because there is an equilibrium between free oxygen and oxygen bound to heme iron. Suppose a solution of myoglobin is placed in a vessel constructed so that a large volume of gas can be mixed with it and so that its color can also be measured through a spectroscope. Without oxygen only the purple color of deoxymyoglobin is observed. If a little oxygen is injected, some of the oxygen combines with some of the deoxymyoglobin to form oxymyoglobin, which is scarlet. The spectroscope measures the proportion of oxymyoglobin in the solution. The injection of oxygen and the spectroscopic measurements are repeated until all the myoglobin has turned scarlet. The results are plotted on a graph with the partial pressure of oxygen on the horizontal axis and the percentage of oxymyoglobin on the vertical axis. The graph has the shape of a rectangular hyperbola: it is steep at the start, when all the myoglobin molecules are free, and it flattens out at the end, when free myoglobin molecules have become so scarce that only a high pressure of oxygen can saturate them.

To understand this equilibrium one must visualize its dynamics. Under the influence of heat the molecules in the solution and in the gas are whizzing around erratically and are constantly colliding. Oxygen molecules are

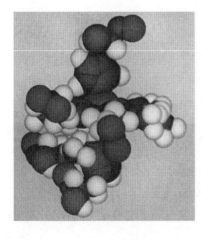

HEME GROUP is the active center of the hemoglobin molecule, the binding site for oxygen. The heme is a flat ring, called a porphyrin, with an iron atom at its center; it is seen here edge on and extending horizontally across the middle of the illustration. Three of the 16 amino acid residues of the globin that are in contact with the heme are also shown. In this computer-generated image each atom is represented by a sphere into which no other atom can penetrate unless the atoms are chemically bonded; where two atoms are bonded the spheres overlap. Carbon atoms are black, nitrogen atoms blue, oxygen atoms red, hydrogen atoms white and the iron atom is rust-colored. The model shows the deoxygenated heme; oxygen binds to the lower side of the iron atom. The picture was generated by Richard J. Feldmann and Thomas K. Porter of the National Institutes of Health from atomic coordinates determined by Giulio Fermi of the Medical Research Council Laboratory of Molecular Biology at Cambridge in England.

entering and leaving the solution, forming bonds with myoglobin molecules and breaking away from them. The number of iron-oxygen bonds that break in one second is proportional to the number of oxymyoglobin molecules. The number of bonds that form in one second is proportional to the frequency of collisions between myoglobin and oxygen, which is determined in turn by the product of their concentrations. When more oxygen is added to the gas, more oxygen molecules dissolve, collide with and bind to myoglobin; this raises the number of oxymyoglobin molecules present and therefore also the number of iron-oxygen bonds liable to break, until the number of myoglobin molecules combining with oxygen in one second becomes equal to the number that lose their oxygen in one second. When that happens, a chemical equilibrium has been established.

The equilibrium is best represented by a graph in which the logarithm of the ratio of oxymyoglobin molecules ($Y$) to deoxymyoglobin molecules ($1 - Y$) is plotted against the logarithm of the partial pressure of oxygen. The hyperbola now becomes a straight line at 45 degrees to the axes. The intercept of the line with the horizontal axis drawn at $Y/(1 - Y) = 1$ gives the equilibrium constant $K$. This is the partial pressure of oxygen at which exactly half of the myoglobin molecules have taken up oxygen. The greater the affinity of the protein for oxygen, the lower the pressure needed to achieve half-saturation and the smaller the equilibrium constant. The 45-degree slope remains unchanged, but lower oxygen affinity shifts the line to the right and higher affinity shifts it to the left.

If the same experiment is done with blood or with a solution of hemoglobin, an entirely different result is obtained. The curve rises gently at first, then steepens and finally flattens out as it approaches themyoglobin curve. This strange sigmoid shape signifies that oxygen-free molecules (deoxyhemoglobin) are reluctant to take up the first oxygen molecule but that their appetite for oxygen grows with the eating. Conversely, the loss of oxygen by some of the hemes lowers the oxygen affinity of the remainder. The distribution of oxygen among the hemoglobin molecules in a solution therefore follows the biblical parable of the rich and the poor: "For unto every one that hath shall be given. and he shall have abundance: but from him that hath not shall be taken away even that which he hath." This phenomenon suggests there is some kind of communication between the hemes in each molecule, and physiologists have therefore called it heme-heme interaction.

A better picture of the underlying mechanism of heme-heme interaction is obtained in a logarithmic graph. The equilibrium curve then begins with a straight line at 45 degrees to the axes, because at first oxygen molecules are so scarce that only one heme in each hemoglobin molecule has a chance of catching one of them, and all the hemes therefore react independently, as in myoglobin. As more oxygen flows in the four hemes in each molecule begin to interact and the curve steepens. The tangent to its maximum slope is known as Hill's coefficient (n), after the physiologist A. V. Hill, who first attempted a mathematical analysis of the oxygen equilibrium. The normal value of Hill's coefficient is about 3; without heme-heme interaction it becomes unity. The curve ends with another line at 45 degrees to the axes because oxygen has now become so abundant that only the last heme in each molecule is likely to be free, and all the hemes in the solution react independently once more.

## Cooperative Effects

Hill's coefficient and the oxygen affinity of hemoglobin depend on the concentration of several chemical factors in the red blood cell: protons (hydrogen atoms without electrons, whose concentration can be measured as pH), carbon dioxide ($CO_2$), chloride ions ($Cl^-$) and a compound of glyceric acid and phosphate called 2,3-diphosphoglycerate (DPG). Increasing the concentration of any of these factors shifts the oxygen equilibrium curve to the right, toward lower oxygen affinity, and makes it more sigmoid. Increased temperature also shifts the curve to the right, but it makes it less sigmoid. Strangely, none of these factors. with the exception of temperature. influences the oxygen equilibrium curve of myoglobin, even though the chemistry and structure of myoglobin are related closely to those of the individual chains of hemoglobin.

What is the purpose of these extraordinary effects? Why is it not good enough for the red cell to contain a simple oxygen carrier such as myoglobin?

Such a carrier would not allow enough of the oxygen in the red cell to be unloaded to the tissues, nor would it allow enough carbon dioxide to be carried to the lungs by the blood plasma. The partial pressure of oxygen in the lungs is about 100 millimeters of mercury, which is sufficient to saturate hemoglobin with oxygen whether the equilibrium curve is sigmoid or hyperbolic. In venous blood the pressure is about 35 millimeters of mercury; if the curve were hyperbolic, less than 10 percent of the oxygen carried would be released at that pressure, so that a man would asphyxiate even if he breathed normally.

The more pronounced the sigmoid shape of the equilibrium curve is the greater the fraction of oxygen that can be released. Several factors conspire to that purpose. Oxidation of nutrients by the tissues liberates lactic acid and carbonic acid; these acids in turn liberate protons, which shift the curve to the right, toward lower oxygen affinity, and make it more sigmoid. Another important regulator of the oxygen affinity is DPG. The number of DPG molecules in the red cell is about the same as the number of hemoglobin molecules, 280 million, and probably remains fairly constant during circulation; a shortage of oxygen, however, causes more DPG to be made, which helps to release more oxygen. With a typical sigmoid curve nearly half of the oxygen carried can be released to the tissues. The human

SUBUNIT OF HEMOGLOBIN consists of a heme group enfolded in a polypeptide chain. The polypeptide is a linear sequence of amino acid residues, each of which is represented here by a single dot, marking the position of the central (alpha) carbon atom. The chain begins with an amino group ($NH_3$) and ends with a carboxyl group (COOH). Most of the polypeptide is coiled up to form helical segments but there are also nonhelical regions. The computer-generated diagram of a horse-hemoglobin subunit was prepared by Feldmann and Porter.

fetus has a hemoglobin with the same alpha chains as the hemoglobin of the human adult but different beta chains, resulting in a lower affinity for DPG. This gives fetal hemoglobin a higher oxygen affinity and facilitates the transfer of oxygen from the maternal circulation to the fetal circulation.

Carbon monoxide (CO) combines with the heme iron at the same site as oxygen, but its affinity for that site is 150 times greater; carbon monoxide therefore displaces oxygen, which explains why it is so toxic. In heavy smokers up to 20 percent of the oxygen combining sites can be blocked by carbon monoxide, so that less oxygen is carried by the blood. In addition carbon monoxide has an even more sinister effect. The combination of one of the four hemes in any hemoglobin molecule with carbon monoxide raises the oxygen affinity of the remaining three hemes by heme-heme interaction. The oxygen equilibrium curve is therefore shifted to the left, which diminishes the fraction of the oxygen carried that can be released to the tissues.

If protons lower the affinity of hemoglobin for oxygen, then the laws of action and reaction demand that oxygen lower the affinity of hemoglobin for protons. Liberation of oxygen causes hemoglobin to combine with protons and vice versa; about two protons are taken up for every four molecules of oxygen released, and two protons are liberated again when four molecules of oxygen are taken up. This reciprocal action is known as the Bohr effect and is the key to the mechanism of carbon dioxide transport. The carbon dioxide released by respiring tissues is too insoluble to be transported as such, but it can be rendered more soluble by combining with water to form a bicarbonate ion and a proton. The chemical reaction is written

$$CO_2 + H_2O \rightarrow HCO_3^- + H^+$$

In the absence of hemoglobin this reaction would soon be brought to a halt by the excess of protons produced, like a fire going out when the chimney is blocked. Deoxyhemoglobin acts as a buffer, mopping up the protons and tipping the balance toward the formation of soluble bicarbonate. In the lungs the process is reversed. There, as oxygen binds to hemoglobin, protons are cast off, driving carbon dioxide out of solution so that it can be exhaled. The reaction between carbon dioxide and water is catalyzed by carbonic anhydrase, an enzyme in the red cells. The enzyme speeds up the reaction to a rate of about half a million molecules per second, one of the fastest of all known biological reactions.

There is a second but less important mechanism for transporting carbon dioxide. The gas binds more readily to deoxyhemoglobin than it does to oxyhemoglobin, so that it tends to be taken up when oxygen is liberated and cast off when oxygen is bound. The two mechanisms of carbon dioxide transport are antagonistic: for each molecule of carbon dioxide bound to deoxyhemoglobin either one or two protons are released, which oppose the conversion of other molecules of carbon dioxide to bicarbonate. Positively charged protons entering the red cell draw negatively charged chloride ions in with them, and these ions too are bound more readily by deoxyhemoglobin than by oxyhemoglobin. DPG is synthesized in the red cell itself and cannot leak out through the cell membrane. It is strongly bound by deoxyhemoglobin and only very weakly bound by oxyhemoglobin.

Heme-heme interaction and the interplay between oxygen and the other four ligands are known collectively as the cooperative effects of hemoglobin. Their discovery by a succession of able physiologists and biochemists took more than half a century and aroused many controversies. In 1938 Felix Haurowitz of the Charles University in Prague made another vital observation. He discovered that deoxyhemoglobin and oxyhemoglobin form different crystals, as though they were different chemical substances, which implied that hemoglobin is not an oxygen tank but a molecular lung because it changes its structure every time it takes up oxygen or releases it.

## Theory of Allostery

The discovery of an interaction among the four hemes made it obvious that they must be touching. but in science what is obvious is not necessarily true. When the structure of hemoglobin was finally solved, the hemes were found to lie in isolated pockets on the surface of the subunits. Without contact between them how could one of them sense whether the others had combined with oxygen? And how could as heterogeneous a collection of chemical agents as protons, chloride ions, carbon dioxide and diphosphoglycerate influence the oxygen equilibrium curve in a similar way? It did not seem plausible that any of them could bind directly to the hemes, that all of them could bind at any other common site, although there again it turned out we were wrong. To add to the mystery, none of these agents affected the oxygen equilibrium of myoglobin or of isolated subunits of hemoglobin. We now know that all the cooperative effects disappear if the hemoglobin molecule is merely split in half, but this vital clue was missed. Like Agatha Christie, nature kept it to the last to make the story more exciting.

There are two ways out of an impasse in science: to experiment or to think. By temperament, perhaps. I experimented, whereas Jacques Monod thought. In the end our paths converged.

Monod's scientific life had been devoted to finding out what regulates the growth of bacteria. The key to this problem appeared to be regulation of the synthesis and catalytic activity of enzymes. Monod and François Jacob had discovered that the activity of certain enzymes is controlled by switching their synthesis on and off at the gene; they and others then found a second mode of regulation that appeared to operate switches on the enzymes themselves.

In 1965 Monod and Jean-Pierre Changeux of the Pasteur Institute in Paris, together with Jeffries Wyman of the University of Rome, recognized that the enzymes in the latter class have certain features in common with hemoglobin. They are all made of several sub-units, so that each molecule includes several sites with the same catalytic activity, just as hemoglobin includes several hemes that bind oxygen, and they all show similar cooperative effects. Monod and his colleagues knew that deoxyhemoglobin and oxyhemoglobin have different structures, which made them suspect that the enzymes too may exist in two (or at least two) structures. They postulated that these structures should be distinguished by the arrangement of the subunits and by the number and strength of the bonds between them.

If there are only two alternative structures, the one with fewer and weaker bonds between the subunits would be free to develop its full catalytic activity (or oxygen affinity); this structure has therefore been labeled R, for "relaxed." The activity would be damped in the structure with more and stronger bonds between the subunits; this form is called $T$ for "tense." In either of these structures the catalytic activity (or oxygen affinity) of all the subunits in one molecule should always remain equal. This postulate of symmetry allowed the properties of allosteric enzymes to be described by a neat mathematical theory with only three independent variables: $K_R$ and $K_t$ which in hemoglobin denote the oxygen equilibrium constants of the $R$ and $T$ structures respectively, and $L$, which stands for the number of molecules in the $T$ structure divided by the number in the $R$ structure, the ratio being measured in the absence of oxygen. The term allostery (from the Greek roots *allos*, "other," and *stereos*, "solid") was coined because the regulator molecule that switches the activity of the enzyme on or off has a structure different from that of the molecule whose chemical transformation the enzyme catalyzes.

This ingenious theory simplified the interpretation of the cooperative effects enormously. The progressive increase in oxygen affinity illustrated by the parable of the rich and the poor now arises not from any direct interaction between the hemes but from the switchover from the $T$ structure with low affinity to the $R$ structure with high affinity. This transformation should take place either when the second molecule of oxygen is bound or when the third is bound. Chemical agents that do not bind to the hemes might lower the oxygen affinity by biasing the equilibrium between the two structures toward the $T$ form, which would make the transition to the $R$ structure come after, say, three molecules of oxygen have been bound rather than after two molecules have been bound. In terms of allosteric theory such agents would raise $L$, the fraction of molecules in the $T$ structure, without altering the oxygen equilibrium constants $K_t$ and $K_R$ of the two structures.

## Atomic Structures

My own approach to the problem was also influenced by Haurowitz' discovery that oxyhemoglobin and deoxyhemoglobin have different structures. Gradually I came to realize that we would never explain the intricate functions of hemoglobin without solving the structures of both crystal forms at a resolution high enough to reveal atomic detail.

In 1970, 33 years after I had taken my first X-ray-diffraction pictures of hemoglobin, that stage was finally reached. Hilary Muirhead, Joyce M. Baldwin, Gwynne Goaman and I got a good map of the distribution of matter not in oxyhemoglobin but in the closely related methemoglobin of horse, in which the iron is ferric and the place of oxygen is taken by a water molecule. William Bolton and I got a map of horse deoxyhemoglobin, and Muirhead and Jonathan Greer got one of human deoxyhemoglobin. These maps served as guides for the construction of three atomic models, each a jungle of brass spokes and steel connectors supported on brass scaffolding, edifices of labyrinthine complexity nearly four feet in diameter. At first it was hard to see the trees for the forest.

In allosteric terms our methemoglobin model represented the R structure and our two deoxyhemoglobin models the T structure. We scanned them eagerly for clues to the allosteric mechanism but could not see any at first because the general structure of the subunits was similar in all three models. The alpha chains included seven helical segments and the beta chains eight helical segments interrupted by corners and non-helical segments. Each chain enveloped its heme in a deep pocket, which exposed only the edge where two propionic acid side chains of the porphyrin dip into the surrounding water.

The heme makes contact with 16 amino acid side chains from seven segments of the chain. Most of these side chains are hydrocarbons; the two exceptions are the heme-linked histidines, which lie on each side of the heme and play an important part in the binding of oxygen. The side chain of histidine ends in an imidazole ring made of three carbon atoms, two nitrogen atoms and either four or five hydrogen atoms. One of these histidines, called the proximal histidine, forms a chemical bond with the heme iron. The other histidine, called the distal one, lies on the opposite side of the heme, in contact with it and with the bound oxygen but without forming a covalent chemical bond with either. Apart from these histidines, most of the side chains in the interior of the subunits, like those near the hemes, are hydrocarbons. The exterior of the hemoglobin molecule is lined with side chains of all kinds, but electrically charged and dipolar ones predominate. Thus each subunit is waxy inside and soapy outside, which makes it soluble in water but impermeable to it.

The four subunits are arranged at the vertexes of a tetrahedron around a twofold symmetry axis. Since a tetrahedron has six edges, there are six areas of contact between the subunits. The twofold symmetry leaves four distinct contacts, which cover about a fifth of the surface area of the subunits. Sixty percent of that area is made up of the $alpha_1$-$beta_1$ and $alpha_2$-$beta_2$ contacts, each of which includes about 35 amino acid side chains tightly linked by from 17 to 19 hydrogen bonds. [Hydrogen bonds are made between atoms of nitrogen (N) and oxygen (0) through an intermediate hydrogen atom (H), for instance N-H...N. N-H...O. O-H... O or O-H... N. The hydrogen is bonded strongly to the atom on the left and weakly to the one on the right.]

The numerous hydrogen bonds between the $alpha_1$-$beta_1$ and $alpha_2$-$beta_2$ subunits make them cohere so strongly that their contact is hardly altered by the reaction with oxygen, and they move as rigid bodies in the transition between the $T$ and the $R$ structures. On the other hand, the contact alpha-$beta_2$ in the $R$ structure looked quite different from that in the $T$ structure. This contact includes fewer side chains than $alpha_1$-$beta_1$ and is designed so that it acts as a snap-action switch, with two alternative stable positions, each braced by a different set of hydrogen bonds. We wondered at first whether these bonds were stronger and more numerous in the $T$ structure than they are in the $R$ structure, but that did not seem to be the case.

Where, then, were the extra bonds between the subunits in the $T$ structure that allosteric theory demanded? We spotted them at the ends of the polypeptide chains. In the $T$ structure the last amino acid residue of each chain forms salt bridges with neighboring subunits. (A salt bridge is a bond between a nitrogen atom, carrying a positive charge, and an oxygen atom, carrying a negative charge.) In our maps of the $R$ structure the last two residues of each chain were blurred. At first I suspected this to be due to error, but improved maps made by my colleagues Elizabeth Heidner and Robert Ladner have convinced us that the final residues remain invisible because they are no longer tethered and wave about like reeds in the wind.

Geometrically, the transition between the two structures consists of a rock-and-roll movement of the dimer $alpha_1$-$beta_1$ with respect to the dimer alpha$_2$-beta$_2$. Baldwin has shown that if one dimer is held fixed, the movement of the other one can be represented by a rotation of some 15 degrees about a suitably placed axis together with a small shift along the same axis. The movement is brought about by subtle changes in the internal structure of the subunits that accompany the binding and dissociation of oxygen.

## Function of the Salt Bridges

The salt bridges at the ends of the polypeptide chains clearly provide the extra bonds between the subunits in the $T$ structure predicted by Monod, Changeux and Wyman. They also explain the influence on the oxygen equilibrium curve of all the chemical factors that had puzzled us so much. All agents that lower the oxygen affinity do so either by strengthening existing salt bridges in the $T$ structure or by adding new ones. Not all these extra bonds, however, are between the subunits; some are within the subunits and oppose the subtle structural changes the subunits undergo on combination with oxygen.

The salt bridges explain both the lowering of the oxygen affinity by protons and the uptake of protons on release of oxygen. Protons increase the number of nitrogen atoms carrying a positive charge. For example, the imidazole ring of the amino acid histidine can exist in two states, uncharged when only one of its nitrogen atoms carries a proton and positively charged when both do. In neutral solution each histidine has a 50 percent chance of being positively charged. The more acid the solution, or in other words the higher the concentration of protons, the greater the chance of a histidine becoming positively charged and forming a salt bridge with an oxygen atom carrying a negative charge. Conversely, the transition from the $R$ structure to the $T$ structure brings negatively charged oxygen atoms into proximity with an uncharged nitrogen atom and thereby diminishes the work that has to be done to give the nitrogen atom a positive charge. As a result a histidine that has no more than a 50 percent chance of being positively charged in the $R$ structure has a 90 percent chance in the $T$ structure, so that more protons are taken up from the solution by hemoglobin in the $T$ structure.

Hemoglobin includes one other set of groups that behave in this way:

they are the amino groups at the start of the polypeptide chains, but their nitrogen atoms take up protons only if the concentration of carbon dioxide is low. If it is high, these nitrogens are liable to lose protons and to combine instead with carbon dioxide to form a carbamino compound. The physiologists F. J. W. Roughton and J. K. W. Ferguson proposed in 1934 that this mechanism plays a part in the transport of carbon dioxide, but their proposal was treated with skepticism until it was confirmed 35 years later by my colleague John Kilmartin, working with Luigi Rossi-Bernardi at the University of Milan. I was pleased that Roughton, who had fathered their experiment, was still alive to see his ideas vindicated, My colleague Arthur R. Arnone, now at the University of Iowa, then showed that in the $T$ structure such carbamino groups, which carry a negative charge, form salt bridges with positively charged groups of the globin and are therefore more stable than they are in the $R$ structure. This finding explains why deoxyhemoglobin has a higher affinity for carbon dioxide than oxyhemoglobin and conversely why carbon dioxide lowers the oxygen affinity of hemoglobin.

The positions in hemoglobin taken up by chloride ions are still uncertain. Arnone has spotted sites in the $T$ structure where other negatively charged ions bind, and these might also be the chloride binding sites. If they are, then chloride ions also brace the $T$ structure by forming additional salt bridges.

The most striking difference between the $T$ and the $R$ structures is the width of the gap between the two beta chains. In the $T$ structure the two chains are widely separated and the opening between them is lined by amino acid side chains carrying positive charges. This opening is tailor-made to fit the molecule of 2.3-diphosphoglycerate and to compensate its negative charges, so that the binding of DPG adds another set of salt bridges to the $T$ structure. In the $R$ structure the gap narrows, and DPG has to drop out.

## The Trigger

How does combination of the heme irons with oxygen make the subunits click from the $T$ structure to the $R$ structure? Compared with the hemoglobin molecule, an oxygen molecule is like the flea that makes the elephant jump. Conversely, how does the *T* structure impede the uptake of oxygen? What difference between the two structures is there at the heme that could bring about a several-hundred-fold change in oxygen affinity?

In oxyhemoglobin the heme iron is bound to six atoms: four nitrogen atoms of the porphyrin, which neutralize the two positive charges of the ferrous iron; one nitrogen atom of the proximal histidine, which links the heme to one of the helical segments of the polypeptide chain (helix *F*), and one of the two atoms of the oxygen molecule. In deoxyhemoglobin the oxygen position remains empty, so that the iron is bound to only five atoms.

I wondered whether the heme pockets might be narrower in the *T* structure than in the *R* structure, so that they had to widen to let the oxygen in. This widening might be geared to break the salt bridges, rather like a childish mechanism. When the atomic model of horse deoxyhemoglobin emerged. Bolton and I saw some truth in this idea because in the beta subunits a side chain of the amino acid valine next to the distal histidine blocked the site that oxygen would have to occupy. The alpha subunits, however, showed no such obstruction. Then we noticed the odd positions of the iron atoms. In methemoglobin, which has the *R* structure, the iron atoms had been displaced very slightly from the porphyrin plane toward the proximal histidine, but in deoxyhemoglobin (with the *T* structure) the displacement stood out as one of the most striking features of our maps. In each subunit the iron atom had carried the proximal histidine and helix *F* with it, so that they too had moved away from the porphyrin plane. I quickly realized that this might be the long-sought trigger.

Recently Arnone and my colleague Lynn Ten Eyck have obtained an excellent map of human deoxyhemoglobin to which Giulio Fermi fitted an atomic model of heme by computer methods. Fermi's calculations show that each iron atom is displaced by $.6(\pm.1)$ angstrom unit from the mean plane of the porphyrin. (One angstrom unit is $10^{-10}$ meter.) The nitrogen atom of the proximal histidine, to which the iron atom is bound, lies at a distance of $2.7(\pm.1)$ angstroms from the same plane. So far we have no direct measure of the corresponding displacements in oxyhemoglobin because oxyhemoglobin oxidizes to methemoglobin in the X-ray beam. There the iron atoms are displaced from the porphyrin plane by 1 angstrom in the alpha subunits and by .2 angstrom in the beta subunits; the corresponding displacements of the histidine nitrogens are 2.2 and 2.4 angstroms. Judging by the structures of model compounds, the displacement of the histidine nitrogen in oxyhemoglobin should be 2.1 angstroms. which means that the nitrogen would be .6 angstrom closer to the porphyrin plane than it is in deoxyhemoglobin. This shift would trigger the transition from the *T* structure to the *R* structure.

How is this movement transmitted to the contacts between the subunits and to the salt bridges? One might as well puzzle out how a cat jumps off a wall from one picture of the cat on the wall and another of it on the ground, because our static models of deoxyhemoglobin and methemoglobin do not show what happens in the transition between the *T* and the *R* structures. I tried a bold guess. The second amino acid residue from the end in each chain is a tyrosine whose side chain carries a phenol group, that is, a benzene ring with a hydroxyl group (OH) attached. In the *T* structure the tyrosine in each subunit is wedged into a pocket between helixes *F* and *H* and its hydroxyl group is tethered by a hydrogen bond to an oxygen atom in the *FG* segment of the main polypeptide chain. In carbonmonoxyhemoglobin, which is the nearest relative of oxyhemoglobin and which has the *R* structure, the tyrosines are free. Hence there must be some mechanism that loosens the tyrosines when oxygen is bound.

As I wondered what this mechanism might be I saw that the movement of the proximal histidine toward the porphyrin plane that accompanies oxygen binding pulls helix *F* in a direction that narrows the pocket into which the tyrosine must fit. If the tyrosine were squeezed out of its pocket, it would tear the last amino acid residue of the chain away from its salt-bridged partner. In this way one salt bridge might be broken for each heme that combined with oxygen in the *T* structure. When enough salt bridges have been broken, the *T* structure would become unstable and click to the *R* structure.

If movement of the proximal histidine and the iron toward the porphyrin puts into motion a set of levers that loosens the tyrosines and breaks the salt bridges, then the making of the bridges and the binding of the tyrosines into their pockets must cause the same set of levers to go into reverse and move the histidine and the iron away from the porphyrin. The oxygen molecule on the other side cannot follow because it bumps against the four porphyrin nitrogen atoms, and so the iron-oxygen bond is stretched until it finally snaps.

To be guided by the atomic models toward the molecular mechanism of respiratory transport seemed like a dream. But was it true? Would the mechanism stand the cold scrutiny of experiment? It has been said that scientists do not pursue the truth, it pursues them.

## Testing the Salt Bridges

According to allosteric theory, there should be no heme-heme interaction without a transition between the $T$ and the $R$ structures. This prediction was also tested by Kilmartin. He cleaved the final amino acid residue from the ends of all four polypeptide chains, so that there should be no salt bridges to stabilize the $T$ structure. This modified hemoglobin maintained the $R$ structure even in the absence of oxygen and showed a hyperbolic oxygen equilibrium curve with high oxygen affinity. Kilmartin then selected an abnormal human hemoglobin that can be made to maintain the T structure even when it is saturated with oxygen. Again the curve was hyperbolic, but it was shifted to lower oxygen affinity, so that the central thesis of allostery was proved.

The next question concerned the exact role of the salt bridges. At one extreme the strain arising from the combination of any one heme with oxygen might be distributed uniformly throughout the molecule, so that there would be no change in oxygen affinity until all the salt bridges broke in unison, when the structure clicked from $T$ to $R$. This would fit pure allosteric theory, according to which the salt bridges should do no more than raise $L$, the fraction of molecules in the $T$ structure. According to my mechanism, on the other hand, a salt bridge should break every time an oxygen combines with the $T$ structure. If that were true, the salt bridges should raise $K_T$ as well as $L$, that is to say, they should also lower the oxygen affinity of the $T$ structure. To the general reader this may seem like a fine distinction, but to the workers in the field it seemed to cut at the roots of the mechanism and raised passionate controversies that are still going on.

Several experimental results favor my version of the mechanism. In 1965 Eraldo Antonini, Todd M. Schuster, Maurizio Brunori and Jeffries Wyman at the University of Rome, and in 1970 R. D. Gray at Cornell University, showed that binding of oxygen and liberation of Bohr protons go hand in hand right from the start, while hemoglobin is still in the $T$ structure. Kilmartin then showed that most of the Bohr protons come from the rupture of salt bridges. Taken together, the two results prove that salt bridges are broken on binding of oxygen by the $T$ structure, which implies that they lower its oxygen affinity. To test whether they really do we had to make an accurate comparison of the oxygen equilibrium curves of normal hemoglobin and a hemoglobin that lacks one of the salt bridges. Kiyohiro Imai and Hideki Morimoto of the University of Osaka had just developed an ingenious method that allows an oxygen equilibrium curve to be measured precisely and fast with only .1 milliliter of hemoglobin solution. Imai came to Cambridge to build one of his new machines and with Kilmartin measured the equilibrium curves of hemoglobins that lacked specific salt bridges. They found that the absence of any of the bridges left $K_R$ unchanged but lowered both $L$ and $K_T$, in accord with my mechanism.

Paradoxically, another set of observations contradicts these findings. My colleagues Leigh Anderson and Kilmartin, together with Seiji Ogawa of Bell Laboratories, have shown that the salt bridges break only if the hemoglobin molecule is free to click to the $R$ structure but not if that transition is stopped. This happens in certain abnormal human hemoglobins and in fish hemoglobins in acid solution, where the $T$ structure is unusually stable. It seems the $T$ structure must be free to bend and stretch so as to shake off its shackles; if it is laced too tightly, it fails to respond.

I have suggested that the transition from the $T$ to the $R$ structure is triggered mainly by the movement of the heme iron toward the porphyrin ring. What makes the iron move? There are two reasons, one steric and the other electronic. If the iron is bound to atoms on both sides of the heme, then their attraction by the iron and repulsion by the porphyrin nitrogens tend to balance the iron in the center of the ring. On the other hand, if the iron is bound only to the proximal histidine while the oxygen site is empty, then repulsion between the porphyrin nitrogens and the histidine is not balanced by repulsion between the porphyrin nitrogens and the oxygen, so that the histidine is pushed away from the porphyrin and pulls the iron with it.

The electronic story is more complex. The ferrous iron atom has six outermost electrons. In oxyhemoglobin these form three pairs located halfway between the bonds that join the iron to its six surrounding atoms. Repulsion between the electrons of the iron and the electrons of the surrounding atoms is thereby minimized. In deoxyhemoglobin, on the other hand, four of the six electrons are unpaired and two of them lie along bond directions. where they repel the surrounding atoms of the porphyrin ring. This repulsion tends to push the iron farther out of the porphyrin plane than the repulsion between the proximal histidine and the porphyrin nitrogen atoms would do on its own.

## Testing the Trigger

Suppose the iron does move in and out of the porphyrin plane every time it combines with or loses a molecule of oxygen. How could we find out if it is really this movement that triggers the allosteric transition between the two structures? I could think of no experiment that would answer the question directly, but I argued that if my proposition were true, then by the laws of action and reaction a forced transition from $R$ to $T$ must put the gears into reverse and pull the iron and the histidine away from the porphyrin ring. In that case the $T$ structure should exercise a tension on the heme. which should be detectable by physical methods. My teacher David Keilin always told me to work with colored proteins because the spectra of the light they absorb can reveal so much. Hemoglobin is doubly blessed because one can feel its pulse both by its absorption spectrum and by the magnetic properties of its iron atoms.

Before we could exploit these properties we had to find a way of switching the structure from $R$ to $T$ other than the usual way of removing the oxygen. Sanford R. Simon of the State University of New York at Stony Brook and I found that this could be done with an analogue of DPG, a substance called inositol hexaphosphate (IHP), which has six phosphate groups in place of the two of DPG and therefore binds to the $T$ structure more strongly.

When IHP was added to oxyhemoglobin, it caused some of the oxygen to be cast off, as was to be expected. I then replaced the oxygen with nitric oxide (NO) because this gas binds to the iron so strongly that the bond, once formed, cannot be broken. When I added IHP to nitric oxide hemoglobin, the structure switched from $R$ to $T$ and the spectrum changed drastically. Analysis of these and other spectral changes told us what had happened: because the strong bond to nitric oxide had held the iron atom tightly to the plane of the porphyrin, the tension exercised by the $T$ structure had snapped the weaker bond between the iron and the proximal histidine instead. Most remarkably, this had happened primarily in the alpha subunits, whose hemes are 35 angstroms away from the phosphate binding site, rather than in the beta subunits, to which the IHP was actually bound. This experiment was done by Kyoshi Nagai, Attila Szabo and me at Cambridge together with John C. Maxwell and Winslow S. Caughey of Colorado State University at Fort Collins. Robert Cassoly of the Institute of Physicochemical Biology in Paris discovered the spectral changes at the same time we did.

Our experiment proved that the tension exists but did not tell us how large it is. To measure it I decided to exploit certain hemoglobin compounds in which the iron atoms are in a state of equilibrium between a weakly and a strongly paramagnetic state. (A paramagnetic substance cannot be permanently magnetized, as metallic iron can, but is drawn into a magnetic field.) At low temperature all the iron atoms are weakly paramagnetic, and the paramagnetism falls as the temperature rises; above a certain temperature the iron atoms begin to oscillate between the two magnetic states, which causes the total paramagnetism to rise as the temperature rises. Today it is known that the bonds between the iron and its surrounding atoms are slightly longer in the strongly paramagnetic state than they are in the weakly paramagnetic one. Therefore if tension in the $T$ structure stretches the bonds to the iron, it should make the proportion of iron atoms in the strongly paramagnetic state larger in the $T$ structure than it is in the $R$ structure and thereby raise the total paramagnetism of the solution.

After several false starts a lucky coincidence finally brought this experiment off. Robert W. Noble appeared at Cambridge from the State University of New York at Buffalo with his pockets full of carp hemoglobin. He showed me how easily the structure of any of the derivatives of this hemoglobin could be switched from $R$ to $T$ by adding a little acid and IHP. Together we set out for Rome, where Massimo Cerdonio and Calogero Messana had just built a highly sensitive superconducting magnetometer at the Snamprogetti Laboratory, but while changing trains at a London Underground station, I left the thermos with our precious samples on the platform and never saw it again. Luckily we had some more carp hemoglobin in our deep freeze at Cambridge, and with it I started off once more for Rome.

The most useful derivative of carp hemoglobin was a ferric form in which the place of oxygen is taken by an azide ion ($N_3^-$). We measured its paramagnetism in both the $R$ and $T$ structures between –180 and +30 degrees Celsius. The results gave us a tremendous thrill: at all temperatures azide methemoglobin of carp was far more strongly paramagnetic in the $T$ structure than in the $R$ structure, which proved that the $T$ structure does favor the state of the heme with the longer iron-nitrogen bonds. The tension at the heme can be gauged from the difference in energy between the two magnetic equilibriums. My colleague Fermi, with my son Robin Perutz of the University of Oxford, worked out that the difference amounts to about 1,000 calories, a third of the free energy of heme-heme interaction. We are not sure where the remaining two-thirds comes from but suspect that the $R$-to-$T$ transition produces a smaller change of heme structure, and therefore also a smaller change of tension, in azide methemoglobin than uptake and loss of oxygen does.

In the meantime Arieh Warshel of the University of Southern California, Bruce W. Gelin and Martin Karplus of Harvard University, my own colleague Joyce Baldwin and Cyrus Chothia of University College London have tried to disentangle the set of atomic levers that generates the tension in the $T$ structure and relieves it on transition to the $R$ structure. They have demolished some of my early ideas and elaborated others.

All agree that the $T$ structure exerts little or no tension on the deoxygenated heme and that the tension arises only when the iron tries to move toward the porphyrin plane on combination with oxygen, rather as the spring of a screen door is relaxed when it is closed but exerts increasing tension as it is opened. James P. Collman of Stanford University has therefore suggested that one should speak of restraint rather than tension. The restraint may be generated by a lopsided orientation of the proximal histidine with respect to the porphyrin, which brings one of the histidine carbon atoms close to one of the porphyrin nitrogen atoms. Repulsion between these two atoms would restrain the histidine from moving closer to the porphyrin ring. On transition to the $R$ structure a shift and rotation of the heme in relation to helix $F$ straightens out the histidine, so that it and the iron atom can move toward the porphyrin without restraint.

In the beta subunits a movement of the heme with respect to helix $E$, which carries the distal valine and distal histidine, may be more important. In the $T$ structure the valine blocks the oxygen combining site, but after the shift to the $R$ structure the site is uncovered. We do not yet know how much this bolting and unbolting contributes to the free energy of the heme-heme interaction.

All these mechanisms are consistent with my early ideas, but my suggestion that the movement of the proximal histidine is transmitted to the salt bridges by squeezing the penultimate tyrosine out of its crevice was too simplistic. Instead the hydrogen bond that holds the tyrosine in place may be stretched, but this loosening may not be enough to break the salt bridges. They may be loosened further by small perturbations of the bonds between the subunits that have so far eluded analysis.

One of the strangest features of both the $T$ and the $R$ structures is the absence of any entrance to the heme pocket wide enough to allow an oxygen molecule to pass. Either the distal histidine or some other group must swing out of the way, but we do not know how this is done because our X-ray analyses portray static structures, which allow us only to guess at the dynamics of the molecule.

John J. Hopfield of Princeton University once said that hemoglobin plays the same role in biochemistry that the hydrogen atom does in physics, because it serves as a touchstone for new theories and experimental techniques. Hemoglobin is the prototype of protein molecules that change their structure in response to chemical stimuli. Scientists will therefore continue to explore its many-faceted behavior. The mechanism I have outlined here will need further refinement before it can explain all their observations, but I am pleased that its main features have stood up to experimental tests and that it accounts reasonably well for the physiological properties of hemoglobin. I have not mentioned here that it also explains the symptoms of patients who have inherited abnormal hemoglobins, because that is another story. I hope that understanding of the structure and mechanism of the hemoglobin molecule will eventually help to alleviate those symptoms and to interpret the behavior of more complex biological systems.

## ABOUT THE AUTHOR

**M. F. Perutz** (1914-2002) was awarded the Nobel Prize in Physiology or Medicine in 1962 along with John C. Kendrew for their studies of the structures of globular proteins.

# The Hemoglobin Molecule

*Its 10,000 atoms are assembled into four chains,
each a helix with several bends. The molecule has
one shape when ferrying oxygen molecules and
a slightly different shape when it is not.*

## M. F. Perutz

In 1937, a year after I entered the University of Cambridge as a graduate
student, I chose the X-ray analysis of hemoglobin, the oxygen-bearing
protein of the blood, as the subject of my research. Fortunately the
examiners of my doctoral thesis did not insist on a determination of the
structure, otherwise I should have had to remain a graduate student for 23
years. In fact, the complete solution of the problem, down to the location of
each atom in this giant molecule, is still outstanding, but the structure has
now been mapped in enough detail to reveal the intricate three-dimensional
folding of each of its four component chains of amino acid units, and the
positions of the four pigment groups that carry the oxygen-combining sites.

The folding of the four chains in hemoglobin turns out to be closely similar to that of the single chain of myoglobin, an oxygen-bearing protein in muscle whose structure has been elucidated in atomic detail by my colleague John C. Kendrew and his collaborators. Correlation of the structure of the two proteins allows us to specify quite accurately, by purely physical methods, where each amino acid unit in hemoglobin lies with respect to the twists and turns of its chains.

Physical methods alone, however, do not yet permit us to decide which of the 20 different kinds of amino acid units occupies any particular site. This knowledge has been supplied by chemical analysis; workers in the U.S. and in Germany have determined the sequence of the 140-odd amino acid units along each of the hemoglobin chains. The combined results of the two different methods of approach now provide an accurate picture of many facets of the hemoglobin molecule.

In its behavior hemoglobin does not resemble an oxygen tank so much as a molecular lung. Two of its four chains shift back and forth, so that the gap between them becomes narrower when oxygen molecules are bound to the hemoglobin, and wider when the oxygen is released. Evidence that the chemical activities of hemoglobin and other proteins are accompanied by structural changes had been discovered before, but this is the first time that the nature of such a change has been directly demonstrated. Hemoglobin's change of shape makes me think of it as a breathing molecule, but paradoxically it expands, not when oxygen is taken up but when it is released.

When I began my postgraduate work in 1936 I was influenced by three inspiring teachers. Sir Frederick Gowland Hopkins, who had received a Nobel prize in 1929 for discovering the growth-stimulating effect of vitamins, drew our attention to the central role played by enzymes in catalyzing chemical reactions in the living cell. The few enzymes isolated at that time had all proved to be proteins. David Keilin, the discoverer of several of the enzymes that catalyze the processes of respiration, told us how the chemical affinities and catalytic properties of iron atoms were altered when the iron combined with different proteins. J. D. Bernal, the X-ray crystallographer, was my research supervisor. He and Dorothy Crowfoot Hodgkin had taken the first X-ray diffraction pictures of crystals of protein a year or two before I arrived, and they had discovered that protein molecules, in spite of their large size, have highly ordered structures. The wealth of sharp X-ray diffraction spots produced by a single crystal of an enzyme such as pepsin could be explained only if every one, or almost every one, of the 5,000 atoms in the pepsin molecule occupied a definite position that was repeated in every one of the myriad of pepsin molecules packed in the crystal. The notion is commonplace now, but it caused a sensation at a time when proteins were still widely regarded as "colloids" of indefinite structure.

In the late 1930's the importance of the nucleic acids had yet to be discovered; according to everything I had learned the "secret of life" appeared to be concealed in the structure of proteins. Of all the methods available in chemistry and physics, X-ray crystallography seemed to offer the only chance, albeit an extremely remote one, of determining that structure.

The number of crystalline proteins then available was probably not more than a dozen, and hemoglobin was an obvious candidate for study because of its supreme physiological importance, its ample supply and the ease with which it could be crystallized. All the same, when I chose the X-ray analysis of hemoglobin as the subject of my Ph.D. thesis, my fellow students regarded me with a pitying smile. The most complex organic substance whose structure had yet been determined by X-ray analysis was the molecule of the dye phthalocyanin, which contains 58 atoms. How could I hope to locate the thousands of atoms in the molecule of hemoglobin?

## The Function of Hemoglobin

Hemoglobin is the main component of the red blood cells, which carry oxygen from the lungs through the arteries to the tissues and help to carry carbon dioxide through the veins back to the lungs. A single red blood cell contains about 280 million molecules of hemoglobin. Each molecule has 64,500 times the weight of a hydrogen atom and is made up of about 10,000 atoms of hydrogen, carbon, nitrogen, oxygen and sulfur, plus four atoms of iron, which are more important than all the rest. Each iron atom lies at the center of the group of atoms that form the pigment called heme, which gives blood its red color and its ability to combine with oxygen. Each heme group is enfolded in one of the four chains of amino acid units that collectively constitute the protein part of the molecule, which is called globin. The four chains of globin consist of two identical pairs. The members of one pair are known as alpha chains and those of the other as beta chains. Together the four chains contain a total of 574 amino acid units.

In the absence of an oxygen carrier a liter of arterial blood at body temperature could dissolve and transport no more than three milliliters of oxygen. The presence of hemoglobin increases this quantity 70 times. Without hemoglobin large animals could not get enough oxygen to exist. Similarly, hemoglobin is responsible for carrying more than 90 percent of the carbon dioxide transported by venous blood.

Each of the four atoms of iron in the hemoglobin molecule can take up one molecule (two atoms) of oxygen. The reaction is reversible in the sense that oxygen is taken up where it is plentiful, as in the lungs, and released where it is scarce, as in the tissues. The reaction is accompanied by a change in color: hemoglobin containing oxygen, known as oxyhemoglobin, makes arterial blood look scarlet; reduced, or oxygen-free, hemoglobin makes venous blood look purple. The term "reduced" for the oxygen-free form is really a misnomer because "reduced" means to the chemist that electrons have been added to an atom or a group of atoms. Actually, as James B. Conant of Harvard University demonstrated in 1923, the iron atoms in both reduced hemoglobin and oxyhemoglobin are in the same electronic condition: the divalent, or ferrous, state. They become oxidized to the trivalent, or ferric, state if hemoglobin is treated with a ferricyanide or removed from the red cells and exposed to the air for a considerable time; oxidation also occurs in certain blood diseases. Under these conditions hemoglobin turns brown and is known as methemoglobin, or ferrihemoglobin.

Ferrous iron acquires its capacity for binding molecular oxygen only through its combination with heme and globin. Heme alone will not bind oxygen, but the specific chemical environment of the globin makes the combination possible. In association with other proteins, such as those of the enzymes peroxidase and catalase, the same heme group can exhibit quite different chemical characteristics.

The function of the globin, however, goes further. It enables the four iron atoms within each molecule to interact in a physiologically advantageous manner. The combination of any three of the iron atoms with oxygen accelerates the combination with oxygen of the fourth; similarly, the release of oxygen by three of the iron atoms makes the fourth cast off its oxygen faster. By tending to make each hemoglobin molecule carry either four molecules of oxygen or none, this interaction ensures efficient oxygen transport.

I have mentioned that hemoglobin also plays an important part in bearing carbon dioxide from the tissues back to the lungs. This gas is not borne by the iron atoms, and only part of it is bound directly to the globin; most of it is taken up by the red cells and the noncellular fluid of the blood in the form of bicarbonate. The transport of bicarbonate is facilitated by the disappearance of an acid group from hemoglobin for each molecule of oxygen discharged. The reappearance of the acid group when oxygen is taken up again in the lungs sets in motion a series of chemical reactions that leads to the discharge of carbon dioxide. Conversely, the presence of bicarbonate and lactic acid in the tissues accelerates the liberation of oxygen.

Breathing seems so simple, yet it appears as if this elementary manifestation of life owes its existence to the interplay of many kinds of atoms in a giant molecule of vast complexity. Elucidating the structure of the molecule should tell us not only what the molecule looks like but also how it works.

## The Principles of X-Ray Analysis

The X-ray study of proteins is sometimes regarded as an abstruse subject comprehensible only to specialists, but the basic ideas underlying our work are so simple that some physicists find them boring. Crystals of hemoglobin and other proteins contain much water and, like living tissues, they tend to lose their regularly ordered structure on drying. To preserve this order during X-ray analysis crystals are mounted wet in small glass capillaries. A single crystal is then illuminated by a narrow beam of X rays that are essentially all of one wavelength. If the crystal is kept stationary, a photographic film placed behind it will often exhibit a pattern of spots lying on ellipses, but if the crystal is rotated in certain ways, the spots can be made to appear at the corners of a regular lattice that is related to the arrangement of the molecules in the crystal. Moreover, each spot has a characteristic intensity that is determined in part by the arrangement of atoms inside the molecules. The reason for the different intensities is best explained in the words of W. L. Bragg, who founded X-ray analysis in 1913—the year after Max von Laue had discovered that X rays are diffracted by crystals—and who later succeeded Lord Rutherford as Cavendish Professor of Physics at Cambridge:

"It is well known that the form of the lines ruled on a [diffraction] grating has an influence on the relative intensity of the spectra which it yields. Some spectra may be enhanced, or reduced, in intensity as compared with others. Indeed, gratings are sometimes ruled in such a way that most of the energy is thrown into those spectra which it is most desirable to examine. The form of the line on the grating does not influence the positions of the spectra, which depend on the number of lines to the centimetre, but the individual lines scatter more light in some directions than others, and this enhances the spectra which lie in those directions.

"The structure of the group of atoms which composes the unit of the crystal grating influences the strength of the various reflexions in exactly the same way. The rays are diffracted by the electrons grouped around the centre of each atom. In some directions the atoms conspire to give a strong scattered beam, in others their effects almost annul each other by interference. The exact arrangement of the atoms is to be deduced by comparing the strength of the reflexions from different faces and in different orders."

Thus there should be a way of reversing the process of diffraction, of proceeding backward from the diffraction pattern to an image of the arrangement of atoms in the crystal. Such an image can actually be produced, somewhat laboriously, as follows. It will be noted that spots on opposite sides of the center of an X-ray picture have the same degree of intensity. With the aid of a simple optical device each symmetrically related pair of spots can be made to generate a set of diffraction fringes, with an amplitude proportional to the square root of the intensity of the spots. The device, which was invented by Bragg and later developed by H. Lipson and C. A. Taylor at the Manchester College of Science and Technology, consists of a point source of monochromatic light, a pair of plane-convex lenses and a

INTERPRETATION OF X-RAY IMAGE can be done with a special optical device to generate a set of diffraction fringes (*right*) from the spots in an X-ray image (*left*). Each pair of symmetrically related spots produces a unique set of fringes. Thus the spots indexed 2,2 and 2,2 yield the fringes indexed 2,2. A two-dimensional image of the atomic structure of a crystal can he generated by printing each set of fringes on the same sheet of photographic paper. But the phase problem must he solved first.

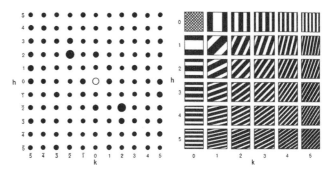

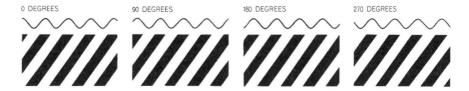

PHASE PROBLEM arises because the spots in an X-ray image do not indicate how the fringes are related in phase to an arbitrarily chosen common origin. Here four identical sets of fringes are related by different phases to the point of origin at the top left corner. The phase marks the distance of the wave crest from the origin, measured in degrees. One wavelength is 360 degrees.

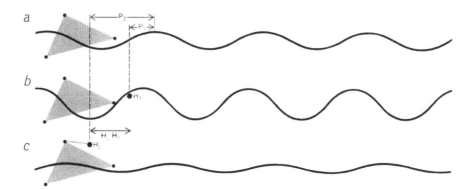

HEAVY-ATOM REPLACEMENT METHOD provides information about phases by changing the intensities of the X ray diffraction pattern. In *a* a highly oversimplified protein (a triangle of three atoms) scatters a sinusoidal wave that represents the amplitude and phase of a single set of fringes. In *b* and *c*, after heavy atoms $H_1$ and $H_2$ are attached to the protein in different positions, the wave is changed in amplitude and phase. The heavy atoms can serve as points of common origin for measuring the magnitude of the phases ($P_1$ and $P_2$) of waves scattered by the unaltered protein. The distance between $H_1$ and $H_2$ must be accurately known.

microscope. The pair of spots in the diffraction pattern is represented by a pair of holes in a black mask that is placed between the two lenses. If the point source is placed at the focus of one of the lenses, the waves of parallel light emerging from the two holes will interfere with one another at the focus of the second lens, and their interference pattern, or diffraction pattern, can be observed or photographed through the microscope.

Imagine that each pair of symmetrically related spots in the X-ray picture is in turn represented by a pair of holes in a mask, and that its diffraction fringes are photographed. Each set of fringes will then be at right angles to the line joining the two holes, and the distance between the fringes will be inversely proportional to the distance between the holes. If the spots are numbered from the center along two mutually perpendicular lines by the indices $h$ and $k$, the relation between any pair of spots and its corresponding set of fringes would be as shown in the illustration.

## The Phase Problem

An image of the atomic structure of the crystal can be generated by printing each set of fringes in turn on the same sheet of photographic paper, or by superposing all the fringes and making a print of the light transmitted through them. At this point, however, a fatal complication arises. In order to obtain the right image one would have to place each set of fringes correctly with respect to some arbitrarily chosen common origin. At this origin the amplitude of any particular set of fringes may show a crest or trough or some intermediate value. The distance of the wave crest from the origin is called the phase. It is almost true to say that by superposing sets of fringes of given amplitude one can generate an infinite number of different images, depending on the choice of phase for each set of fringes. By itself the X-ray picture tells us only about the amplitudes and nothing about the phases of the fringes to be generated by each pair of spots, which means that half the information needed for the production of the image is missing.

The missing information makes the diffraction pattern of a crystal like a hieroglyphic without a key. Having spent years hopefully measuring the intensities of several thousand spots in the diffraction pattern of hemoglobin, I found myself in the tantalizing position of an explorer with a collection of tablets engraved in an unknown script. For some time Bragg and I tried to develop methods for deciphering the phases, but with only limited success. The solution finally came in 1953, when I discovered that a method that had been developed by crystallographers for solving the phase problem in simpler structures could also be applied to proteins.

In this method the molecule of the compound under study is modified slightly by attaching heavy atoms such as those of mercury to definite positions in its structure. The presence of a heavy atom produces marked changes

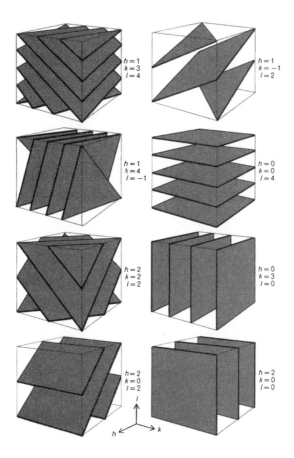

**THREE-DIMENSIONAL FRINGES** are needed to build up an image of protein molecules. For this purpose many different X-ray diffraction images are prepared and symmetrically related pairs of spots are indexed in three dimensions: $h$, $k$ and $l$ and $\bar{h}$, $\bar{k}$ and $\bar{l}$. Each pair of spots yields a three-dimensional fringe like those shown here. Fringes from thousands of spots must be superposed in proper phase to build up an image of the molecule.

in the intensities of the diffraction pattern, and this makes it possible to gather information about the phases. From the difference in amplitude in the absence or presence of a heavy atom, the distance of the wave crest from the heavy atom can be determined for each set of fringes. Thus with the heavy atom serving as a common origin the magnitude of the phase can be measured. The illustration shows how the phase of a single set of fringes, represented by a sinusoidal wave that is supposedly scattered by the oversimplified protein molecule, can be measured from the increase in amplitude produced by the heavy atom $H_1$.

Unfortunately this still leaves an ambiguity of sign; the experiment does not tell us whether the phase is to be measured from the heavy atom in the forward or the backward direction. If $n$ is the number of diffracted spots, an ambiguity of sign in each set of fringes would lead to $2^n$ alternative images of the structure. The Dutch crystallographer J. M. Bijvoet had pointed out some years earlier in another context that the ambiguity could be resolved by examining the diffraction pattern from a second heavy-atom compound.

The heavy atom $H_2$, which is attached to the protein in a position different from that of $H_1$, diminishes the amplitude of the wave scattered by the protein. The degree of attenuation allows us to measure the distance of the wave crest from $H_2$. It can now be seen that the wave crest must be in front of $H_1$; otherwise its distance from $H_1$ could not be reconciled with its distance from $H_2$. The final answer depends on knowing the length and direction of the line joining $H_2$ to $H_1$. These quantities are best calculated by a method that does not easily lend itself to exposition in nonmathematical language. It was devised by my colleague Michael C. Rossmann.

The heavy-atom method can be applied to hemoglobin by attaching mercury atoms to the sulfur atoms of the amino acid cysteine. The method works, however, only if this attachment leaves the structure of the hemoglobin molecules and their arrangement in the crystal unaltered. When I first tried it, I was not at all sure that these stringent demands would be fulfilled, and as I developed my first X-ray photograph of mercury hemoglobin my mood alternated between sanguine hopes of immediate success and desperate forebodings of all the possible causes of failure. When the diffraction spots appeared in exactly the same position as in the mercury-free protein but with slightly altered intensities, just as I had hoped, I rushed off to Bragg's room in jubilant excitement, expecting that the structure of hemoglobin and of many other proteins would soon be determined, Bragg shared my excitement, and luckily neither of us anticipated the formidable technical difficulties that were to hold us up for another five years.

## Resolution of the Image

Having solved the phase problem, at least in principle, we were confronted with the task of building up a structural image from our X-ray data. In simpler structures atomic positions can often be found from representations of the structure projected on two mutually perpendicular planes, but in proteins a three-dimensional image is essential. This can be attained by making use of the three-dimensional nature of the diffraction pattern, The X-ray diffraction pattern can be regarded as a section through a sphere that is filled with layer after layer of diffraction spots. Each pair of spots can be made to generate a set of three-dimensional fringes. When their phases have been measured, they can be superposed by calculation to build up a three-dimensional image of the protein. The final image is represented by a series of sections through the molecule, rather like a set of microtome sections through a piece of tissue, only on a scale 1,000 times smaller.

The resolution of the image is roughly equal to the shortest wavelength of the fringes used in building it up. This means that the resolution increases with the number of diffracted spots included in the calculation. If the image is built up from part of the diffraction pattern only, the resolution is impaired.

In the X-ray diffraction patterns of protein crystals the number of spots runs into tens of thousands. In order to determine the phase of each spot accurately, its intensity (or blackness) must be measured accurately several times over: in the diffraction pattern from a crystal of the pure protein and in the patterns from crystals of several compounds of the protein, each with heavy atoms attached to different positions in the molecule. Then the results have to be corrected by various geometric factors before they are finally used to build up an image through the superposition of tens of thousands of fringes. In the final calculation tens of millions of numbers may have to be added or subtracted. Such a task would have been quite impossible before the advent of high-speed computers, and we have been fortunate in that the development of computers has kept pace with the expanding needs of our X-ray analyses.

While I battled with technical difficulties of various sorts, my colleague John Kendrew successfully applied the heavy-atom method to myoglobin, a protein closely related to hemoglobin [see "The Three-dimensional Structure of a Protein Molecule," by John C. Kendrew; SCIENTIFIC AMERICAN, December, 1961]. Myoglobin is simpler than hemoglobin because it consists of only one chain of amino acid units and one heme group, which binds a single molecule of oxygen. The complex interaction phenomena involved in hemoglobin's dual function as a carrier of oxygen and of carbon dioxide do not occur in myoglobin, which acts simply as an oxygen store.

Together with Howard M. Dintzis and C. Bodo, Kendrew was brilliantly successful in managing to prepare as many as five different crystalline heavy-atom compounds of myoglobin, which meant that the phases of the diffraction spots could be established very accurately. He also pioneered the use of high-speed computers in X-ray analysis. In 1957 he and his colleagues obtained the first three-dimensional representation of myoglobin.

It was a triumph, and yet it brought a tinge of disappointment. Could the search for ultimate truth really have revealed so hideous and visceral-looking an object? Was the nugget of gold a lump of lead? Fortunately, like many other things in nature, myoglobin gains in beauty the closer you look at it. As Kendrew and his colleagues increased the resolution of their X-ray analysis in the years that followed, some of the intrinsic reasons for the molecule's strange shape began to reveal themselves. This shape was found to be not a freak but a fundamental pattern of nature, probably common to myoglobins and hemoglobins throughout the vertebrate kingdom.

In the summer of 1959, nearly 22 years after I had taken the first X-ray pictures of hemoglobin, its structure emerged at last. Michael Rossmann, Ann F. Cullis, Hilary Muirhead, Tony C. T. North and I were able to prepare a three-dimensional electron-density map of hemoglobin at a resolution of 5.5 angstrom units, about the same as that obtained for the first structure of myoglobin two years earlier. This resolution is sufficient to reveal the shape of the chain forming the backbone of a protein molecule but not to show the position of individual amino acids.

As soon as the numbers printed by the computer had been plotted on contour maps we realized that each of the four chains of hemoglobin had a shape closely resembling that of the single chain of myoglobin. The beta chain and myoglobin look like identical twins, and the alpha chains differ from them merely by a shortcut across one small loop.

Kendrew's myoglobin had been extracted from the muscle of the sperm whale; the hemoglobin we used came from the blood of horses. More recent observations indicate that the myoglobins of the seal and the horse, and the hemoglobins of man and cattle, all have the same structure. It seems as though the apparently haphazard and irregular folding of the chain is a pattern specifically devised for holding a heme group in place and for enabling it to carry oxygen.

What is it that makes the chain take up this strange configuration? The extension of Kendrew's analysis to a higher resolution shows that the chain of myoglobin consists of a succession of helical segments interrupted by corners and irregular regions. The helical segments have the geometry of the alpha helix predicted in 1951 by Linus Pauling and Robert B. Corey of the California Institute of Technology. The heme group lies embedded in a fold of the chain, so that only its two acid groups protrude at the surface and are in contact with the surrounding water. Its iron atom is linked to a nitrogen atom of the amino acid histidine.

I have recently built models of the alpha and beta chains of hemoglobin and found that they follow an atomic pattern very similar to that of myoglobin. If two protein chains look the same, one would expect them to have much the same composition. In the language of protein chemistry this implies that in the myoglobins and hemoglobins of all vertebrates the 20 different kinds of amino acid should be present in about the same proportion and arranged in similar sequence.

Enough chemical analyses have been done by now to test whether or not this is true. Starting at the Rockefeller Institute and continuing in our laboratory, Allen B. Edmundson has determined the sequence of amino acid units in the molecule of sperm-whale myoglobin. The sequences of the alpha and beta chains of adult human hemoglobin have been analyzed independently by Gerhardt Braunitzer and his colleagues at the Max Planck Institute for Biochemistry in Munich, and by William H. Konigsberg, Robert J. Hill and their associates at the Rockefeller Institute. Fetal hemoglobin, a variant of the human adult form, contains a chain known as gamma, which is closely related to the beta chain. Its complete sequence has been analyzed by Walter A. Schroeder and his colleagues at the California Institute of Technology. The sequences of several other species of hemoglobin and that of human myoglobin have been partially elucidated.

The sequence of amino acid units in proteins is genetically determined, and changes arise as a result of mutation. Sickle-cell anemia, for instance, is an inherited disease due to a mutation in one of the hemoglobin genes. The mutation causes the replacement of a single amino acid unit in each of the beta chains. (The glutamic acid unit normally present at position No. 6 is replaced by a valine unit.) On the molecular scale evolution is thought to involve a succession of such mutations, altering the structure of protein molecules one amino acid unit at a time. Consequently when the hemoglobins of different species are compared, we should expect the sequences

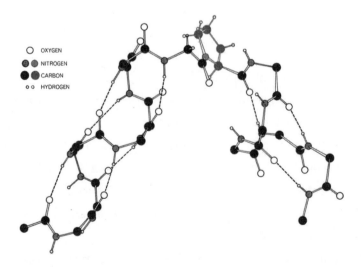

OXYGEN
NITROGEN
CARBON
HYDROGEN

CORNER IN HEMOGLOBIN MOLECULE occurs where a subunit of the amino acid praline falls between two helical regions in the beta chain. The chain is shown bare; all hydrogen atoms and amino acid side branches, except for proline, are removed.

in man and apes, which are close together on the evolutionary scale, to be very similar, and those of mammals and fishes, say, to differ more widely. Broadly speaking, this is what is found. What was quite unexpected was the degree of chemical diversity among the amino acid sequences of proteins of similar three-dimensional structure and closely related function. Comparison of the known hemoglobin and myoglobin sequences shows only 15 positions—no more than one in 10—where the same amino acid unit is present in all species. In all the other positions one or more replacements have occurred in the course of evolution.

What mechanism makes these diverse chains fold up in exactly the same way? Does a template force them to take up this configuration, like a mold that forces a car body into shape? Apart from the topological improbability of such a template, all the genetic and physico-chemical evidence speaks against it, suggesting instead that the chain folds up spontaneously to assume one specific structure as the most stable of all possible alternatives.

## Possible Folding Mechanisms

What is it, then, that makes one particular configuration more stable than all others? The only generalization to emerge so far, mainly from the work of Kendrew, Herman C. Watson and myself, concerns the distribution of the so-called polar and nonpolar amino acid units between the surface and the interior of the molecule.

Some of the amino acids, such as glutamic acid and lysine, have side groups of atoms with positive or negative electric charge, which strongly attract the surrounding water. Amino acid side groups such as glutamine or tyrosine, although electrically neutral as a whole, contain atoms of nitrogen or oxygen in which positive and negative charges are sufficiently separated to form dipoles; these also attract water, but not so strongly as the charged groups do. The attraction is due to a separation of charges in the water molecule itself, making it dipolar. By attaching themselves to electrically charged groups, or to other dipolar groups, the water molecules minimize the strength of the electric fields surrounding these groups and stabilize the entire structure by lowering the quantity known as free energy.

The side groups of amino acids such as leucine and phenylalanine, on the other hand, consist only of carbon and hydrogen atoms. Being electrically neutral and only very weakly dipolar, these groups repel water as wax does. The reason for the repulsion is strange and intriguing. Such hydrocarbon groups, as they are called, tend to disturb the haphazard arrangement of the liquid water molecules around them, making it ordered as it is in ice. The increase in order makes the system less stable; in physical terms it leads to a reduction of the quantity known as entropy, which is the measure of the disorder in a system. Thus it is the water molecules' anarchic distaste for the orderly regimentation imposed on them by the hydrocarbon side groups that forces these side groups to turn away from water and to stick to one another.

Our models have taught us that most electrically charged or dipolar side groups lie at the surface of the protein molecule, in contact with water. Non-polar side groups, in general, are either confined to the interior of the molecule or so wedged into crevices on its surface as to have the least contact with water. In the language of physics, the distribution of side groups is of the kind leading to the lowest free energy and the highest entropy of the protein molecules and the water around them. (There is a reduction of entropy due to the orderly folding of the protein chain itself, which makes the system less stable, but this is balanced, at moderate temperatures, by the stabilizing contributions of the other effects just described.) It is too early to say whether these are the only generalizations to be made about the forces that stabilize one particular configuration of the protein chain in preference to all others.

At least one amino acid is known to be a misfit in an alpha helix, forcing the chain to turn a corner wherever the unit occurs. This is proline. There is, however, only one corner in all the hemoglobins and myoglobins where a proline is always found in the same position: position No. 36 in the beta chain and No. 37 in the myoglobin chain. At other corners the appearance of prolines is haphazard and changes from species to species. Elkan R. Blout of the Harvard Medical School finds that certain amino acids such as valine or threonine, if present in large numbers, inhibit the formation of alpha helices, but these do not seem to have a decisive influence in myoglobin and hemoglobin.

Since it is easier to determine the sequence of amino acid units in proteins than to unravel their three-dimensional structure by X rays, it would be useful to be able to predict the structure from the sequence. In principle enough is probably known about the forces between atoms and about the way they tend to arrange themselves to make such predictions feasible. In practice the enormous number of different ways in which a long chain can be twisted still makes the problem one of baffling complexity.

## Assembling the Four Chains

If hemoglobin consisted of four identical chains, a crystallographer would expect them to lie at the corners of a regular tetrahedron. In such an arrangement each chain can be brought into congruence with any of its three neighbors by a rotation of 180 degrees about one of three mutually perpendicular axes of symmetry. Since the alpha and beta chains are chemically different, such perfect symmetry is unattainable, but the actual arrangement comes very close to it. As a first step in the assembly of the molecule two alpha chains are placed near a twofold symmetry axis, so that a rotation of 180 degrees brings one chain into congruence with its partner.

Next the same is done with the two beta chains. One pair, say the alpha chains, is then inverted and placed over the top of the other pair so that the four chains lie at the corners of a tetrahedron. A true twofold symmetry axis now passes vertically through the molecule, and "pseudo-axes" in two directions perpendicular to the first relate the alpha to the beta chains. Thus the arrangement is tetrahedral, but because of the chemical differences between the alpha and beta chains the tetrahedron is not quite regular.

The result is an almost spherical molecule whose exact dimensions are $64 \times 55 \times 50$ angstrom units. It is astonishing to find that four objects as irregular as the alpha and beta chains can fit together so neatly. On formal grounds one would expect a hole to pass through the center of the molecule because chains of amino acid units, being asymmetrical, cannot cross any symmetry axis. Such a hole is in fact found.

The most unexpected feature of the oxyhemoglobin molecule is the way the four heme groups are arranged. On the basis of their chemical interaction one would have expected them to lie close together. Instead each heme group lies in a separate pocket on the surface of the molecule, apparently unaware of the existence of its partners. Seen at the present resolution, therefore, the structure fails to explain one of the most important physiological properties of hemoglobin.

In 1937 Felix Haurowitz, then at the German University of Prague, discovered an important clue to the molecular explanation of hemoglobin's physiological action. He put a suspension of needle-shaped oxyhemoglobin crystals away in the refrigerator. When he took the suspension out some weeks later, the oxygen had been used up by bacterial infection and the scarlet needles had been replaced by hexagonal plates of purple reduced hemoglobin. While Haurowitz observed the crystals under the microscope, oxygen penetrated between the slide and the cover slip, causing the purple plates to dissolve and the scarlet needles of hemoglobin to re-form. This transformation convinced Haurowitz that the reaction of hemoglobin with oxygen must be accompanied by a change in the structure of the hemoglobin molecule. In myoglobin, on the other hand, no evidence for such a change has been detected.

Haurowitz' observation and the enigma posed by the structure of oxyhemoglobin caused me to persuade a graduate student, Hilary Muirhead, to attempt an X-ray analysis at low resolution of the reduced form. For technical reasons human rather than horse hemoglobin was used at first, but we have now found that the reduced hemoglobins of man and the horse have very similar structures, so that the species does not matter here.

Unlike me, Miss Muirhead succeeded in solving the structure of her protein in time for her Ph.D. thesis. When we examined her first electron-density maps, we looked for two kinds of structural change: alterations in the folding of the individual chains and displacements of the chains with respect to each other. We could detect no changes in folding large enough to be sure that they were not due to experimental error. We did discover, however, that a striking displacement of the beta chains had taken place. The gap between them had widened and they had been shifted sideways, increasing the distance between their respective iron atoms from 33.4 to 40.3 angstrom units. The arrangement of the two alpha chains had remained unaltered, as far as we could judge, and the distance between the iron atoms in the beta chains and their nearest neighbors in the alpha chains had also remained the same. It looked as though the two beta chains had slid apart, losing contact with each other and somewhat changing their points of contact with the alpha chains.

F. J. W. Roughton and others at the University of Cambridge suggest that the change to the oxygenated form of hemoglobin takes place after three of the four iron atoms have combined with oxygen. When the change has occurred, the rate of combination of the fourth iron atom with oxygen

| RESIDUE NUMBER | HEMOGLOBIN | | | MYOGLOBIN |
|---|---|---|---|---|
| | ALPHA | BETA | GAMMA | |
| 81 | MET | LEU | LEU | HIS |
| 82 | PRO | LYS | LYS | GLU |
| 83 | ASN | GLY | GLY | ALA |
| 84 | ALA | THR | THR | GLU |
| 85 | LEU | PHE | PHE | LEU |
| 86 | SER | ALA | ALA | LYS |
| 87 | ALA | THR | GLN | PRO |
| 88 | LEU | LEU | LEU | LEU |
| 89 | SER | SER | SER | ALA |
| 90 | ASP | GLU | GLU | GLN |
| 91 | LEU | LEU | LEU | SER |
| 92 | HIS | HIS | HIS | HIS |
| 93 | ALA | CYS | CYS | ALA |
| 94 | HIS | ASP | ASN | THR |
| 95 | LYS | LYS | LYS | LYS |
| 96 | LEU | LEU | LEU | HIS |
| 97 | ARG | HIS | HIS | LYS |
| 98 | VAL | VAL | VAL | ILEU |
| 99 | ASP | ASP | ASP | PRO |
| 100 | PRO | PRO | PRO | ILEU |
| 101 | VAL | GLU | GLU | LYS |
| 102 | ASP | ASN | ASN | TYR |

| | | |
|---|---|---|
| ALA ALANINE | GLY GLYCINE | PRO PROLINE |
| ARG ARGININE | HIS HISTIDINE | SER SERINE |
| ASN ASPARAGINE | ILEU ISOLEUCINE | THR THREONINE |
| ASP ASPARTIC ACID | LEU LEUCINE | TYR TYROSINE |
| CYS CYSTEINE | LYS LYSINE | VAL VALINE |
| GLU GLUTAMIC ACID | PHE PHENYLALANINE | |

AMINO ACID SEQUENCES are shown for corresponding stretches of the alpha and beta chains of hemoglobin from human adults, the gamma chain that replaces the beta chain in fetal human hemoglobin and sperm-whale myoglobin. Colored bars show where the same amino acid units are found either in all four chains or in the first three. Site numbers for the alpha chain and myoglobin are adjusted slightly because they contain a different number of amino acid subunits overall than do the beta and gamma chains. Over their full length of more than 140 subunits the four chains have only 20 amino acid subunits in common.

is speeded up several hundred times. Nothing is known as yet about the atomic mechanism that sets off the displacement of the beta chains, but there is one interesting observation that allows us at least to be sure that the interaction of the iron atoms and the change of structure do not take place unless alpha and beta chains are both present.

Certain anemia patients suffer from a shortage of alpha chains; the beta chains, robbed of their usual partners, group themselves into independent assemblages of four chains. These are known as hemoglobin H and resemble normal hemoglobin in many of their properties. Reinhold Benesch and Ruth E. Benesch of the Columbia University College of Physicians and Surgeons have discovered, however, that the four iron atoms in hemoglobin H do not interact, which led them to predict that the combination of hemoglobin H with oxygen should not be accompanied by a change of structure. Using crystals grown by Helen M. Ranney of the Albert Einstein College of Medicine, Lelio Mazzarella and I verified this prediction. Oxygenated and reduced hemoglobin H both resemble normal human reduced hemoglobin in the arrangement of the four chains.

The rearrangement of the beta chains must be set in motion by a series of atomic displacements starting at or near the iron atoms when they combine with oxygen. Our X-ray analysis has not yet reached the resolution needed to discern these, and it seems that a deeper understanding of this intriguing phenomenon may have to wait until we succeed in working out the structures of reduced hemoglobin and oxyhemoglobin at atomic resolution.

## Allosteric Enzymes

There are many analogies between the chemical activities of hemoglobin and those of enzymes catalyzing chemical reactions in living cells. These analogies lead one to expect that some enzymes may undergo changes of structure on coming into contact with the substances whose reactions they catalyze. One can imagine that the active sites of these enzymes are moving mechanisms rather than static surfaces magically endowed with catalytic properties.

Indirect and tentative evidence suggests that changes of structure involving a rearrangement of subunits like that of the alpha and beta chains of hemoglobin do indeed occur and that they may form the basis of a control mechanism known as feedback inhibition. This is a piece of jargon that biochemistry has borrowed from electrical engineering, meaning nothing more complicated than that you stop being hungry when you have had enough to eat.

Constituents of living matter such as amino acids are built up from simpler substances in a series of small steps, each step being catalyzed by an enzyme that exists specifically for that purpose. Thus a whole series of different enzymes may be needed to make one amino acid. Such a series of enzymes appears to have built-in devices for ensuring the right balance of supply and demand. For example, in the colon bacillus the amino acid isoleucine is made from the amino acid threonine in several steps. The first enzyme in the series has an affinity for threonine: it catalyzes the removal of an amino group from it. H. Edwin Umbarger of the Long Island Biological Association in Cold Spring Harbor, N.Y., discovered that the action of the enzyme is inhibited by isoleucine, the end product of the last enzyme in the series. Jean-Pierre Changeux of the Pasteur Institute later showed that isoleucine acts not, as one might have expected, by blocking the site on the enzyme molecule that would otherwise combine with threonine but probably by combining with a different site on the molecule.

The two sites on the molecule must therefore interact, and Jacques Monod, Changeux and Francois Jacob have suggested that this is brought about by a rearrangement of subunits similar to that which accompanies the reaction of hemoglobin with oxygen. The enzyme is thought to exist in two alternative structural states: a reactive one when the supply of isoleucine has run out and an unreactive one when the supply exceeds demand. The discoverers have coined the name "allosteric" for enzymes of this kind.

The molecules of the enzymes suspected of having allosteric properties are all large ones, as one would expect them be if they are made up of several subunits. This makes their X-ray analysis difficult. It may not be too hard to find out, however, whether or not a change of structure

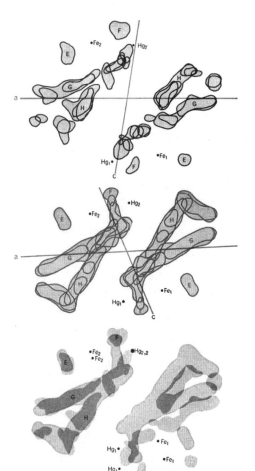

occurs, even if it takes a long time to unravel it in detail. In the meantime hemoglobin will serve as a useful model for the behavior of more complex enzyme systems.

MOVEMENT OF HEMOGLOBIN CHAINS was discovered by comparing portions of the two heta chains in "reduced" (oxygen-free) human hemoglobin (top) with the same portions of horse hemoglobin containing oxygen (middle). The bottom illustration shows the outlines of the top and middle pictures superposed so that the mercury atoms ($Hg_2$) and helical regions (E, F, G, H) of the two chains at left coincide. The iron atoms ($Fe_2$) do not quite match. The chains at right are now seen to he shifted with respect to each other.

## ABOUT THE AUTHOR

**M. F. Perutz** (1914-2002) was awarded the Nobel Prize in Physiology or Medicine in 1962 along with John C. Kendrew for their studies of the structures of globular proteins.

# The Structure of Antibodies

*The basic pattern of the principal class of molecules that neutralize antigens (foreign substances in the body) is four cross-linked chains. This pattern is modified so that antibodies can fit different antigens.*

## R. R. Porter

I t has been known for millenniums that a person who survives a disease such as plague or smallpox is usually able to resist a second infection. Indeed, such immune people were often the only ones available to nurse the sick during severe epidemics. A general understanding of immunity had to await the discovery that microorganisms are the causative agents of infectious disease. Then progress was rapid. A key step was taken in 1890 by Emil Von Behring and Shibasaburo Kitasato, working in the Institute of Robert Koch in Berlin. They showed that an animal could be made immune to tetanus by an injection of the blood serum obtained from an animal that had survived the disease and had developed immunity to it. Serum is the clear fluid that is left behind when a blood clot forms; it contains most of the blood proteins. Thus immunity to tetanus is a function of a substance or substances in the blood. These substances were named antibodies.

Antibodies are produced by all vertebrates as a defense against invasion by certain foreign substances, known collectively as antigens. The most effective antigens are large molecules such as proteins or polysaccharides (and of course the microorganisms that contain these molecules). The demonstration of the appearance of antibodies in the blood is most dramatic if the antigen is a lethal toxin or a pathogenic microorganism: the immune animals live and the non-immune die when injected with the antigen. Innocuous substances such as egg-white protein or the polysaccharide coat of bacteria, however, are equally effective as antigens. The antibodies formed against them can be detected by their ability to combine with antigen, This can be shown in many ways. Perhaps the simplest demonstration is provided by the precipitate that appears in a test tube when a soluble antigen combines with antibody contained in a sample of serum. The most remarkable aspect of this phenomenon is the specificity of the antibody for the antigen injected. That is, the antibody formed will combine only with the antigen injected or with other substances whose structure is closely related.

Numerous different antibodies can be formed. Although an individual

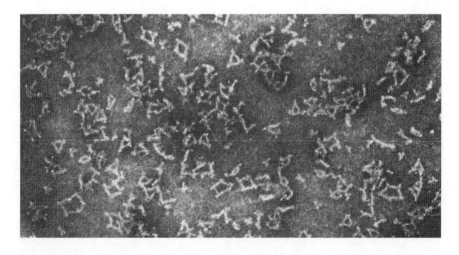

ANTIBODIES BOUND TO ANTIGENS are depicted in this electron micrograph made by Michael Green and Robin Valentine of the National Institute for Medical Research in London. The antigen itself is too small to be visible, but it evidently acts as the coupling agent that binds antibody molecules together to form the various multisided structures. The magnification is about 275,000 diameters.

animal may respond poorly, or perhaps not at all, to a particular antigen, there is no known limit to the number of specific antibodies that one species, for example the rabbit, can synthesize. Conceptually there is a great difference between the capability of one species to synthesize a very large but limited number of antibodies and the capacity to synthesize an infinite number, but an experimental decision as to which is correct is not possible at present.

All antibodies are found in a group of related serum proteins known as immunoglobulins. The challenge to the protein chemist lies in the fact that antibody molecules are surprisingly similar even though they possess an enormous range of specific combining power. Although it is clear that there must be significant differences among antibodies, no chemical or physical property has yet been found that can distinguish between two antibody molecules: one able to combine specifically, say, with an aromatic compound such as a benzene derivative and the other with a sugar, although the benzene compound and the sugar have no common structural features. Antibodies of quite unrelated specificity appear to be identical, within the

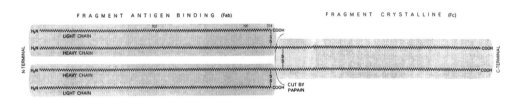

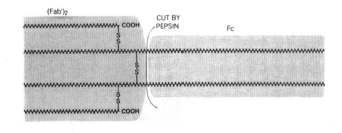

IMMUNOGLOBULIN GAMMA, the chief class of antibody, is a protein molecule consisting of four polypeptide chains held together by disulfide (S—S) bonds. The two light chains are identical, as are the two heavy chains. Depending on the source, the light chains contain from about 210 to 230 amino acid units; the heavy chains vary from about 420 to 440 units. Thus the lengths, the spacing between disulfide bonds and enzyme cleavage points shown here are approximate. The enzyme papain splits the molecule into three fragments: a fragment that forms crystals (Fc) and two fragments (Fob) that do not crystallize but contain the antigen binding sites. Approximately half of each *Fab* fragment is variable in amino acid composition. Site 191 is genetically variable. When immunoglobulin gamma is split by the enzyme pepsin, the *Fab* fragments remain bonded together (Fab')₂ because the cleavage occurs on the other side of the central disulfide bond.

limits of present experimental techniques, except, of course, in their specific combination with antigen.

An antibody can be isolated from the serum of an immunized animal only by using the special property of allowing it to combine with the antigen, freeing the complex from the other serum proteins and then dissociating and separating the antibody and antigen. This can be done by allowing a precipitate to form, washing the precipitate well with salt solution and then suspending the precipitate in weak acid. Under these conditions the antibody-antigen precipitate will dissolve and dissociate, and the antibody and antigen can be separated from each other to yield the purified antibody, As we shall see, however, even this purified material usually contains a variety of antibody molecules that differ slightly in their molecular structure.

If an animal has not been immunized, it will still have a good concentration of immunoglobulin in its blood, usually about 1 percent by weight. This material is believed to be made up of many thousands of different antibodies against microorganisms the animal has encountered during its lifetime or against other antigenic substances that accidentally entered its body. Evidence that this view is correct comes from experiments in which small animals have been born and raised in an entirely germ-free environment. Under these conditions the immunoglobulin content of the blood is much lower, perhaps only 10 percent of the immunoglobulin in the blood of a normal animal, suggesting that mild infections are the main source of antigens.

The immunoglobulins can be isolated from serum by the usual methods of protein separation. Hence the protein chemist has available for study two general kinds of immunoglobulin fraction: a complex mixture of many antibodies and purified antibodies that have been isolated by virtue of their specific affinity for the antigen. It would seem to be a relatively straightforward task, after the great progress made in the techniques of protein chemistry in recent years, to carry out detailed studies of such material and pinpoint the differences. Clearly structural differences responsible for the specific combining power of antibodies must exist among them and should become apparent.

Major difficulties have arisen, however, because the immunoglobulins have been found to be a very complex mixture of molecules and the complexity is not necessarily due to the presence of the many different kinds of antibody. One difficulty is that there are three main classes of immunoglobulins distinguished chemically from one another by size, carbohydrate content and amino acid analysis. Antibodies of any specificity can be found in any of the classes; hence there is no correlation between class and specificity. The class present in the largest amounts in the blood and the most easily isolated is called immunoglobulin gamma. Since most of the work has been done with this material I shall limit my discussion to it.

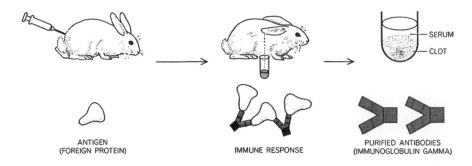

MIXTURE OF SIMILAR ANTIBODIES can be produced by injecting a rabbit or other animal with a purified antigen, typically a large protein of foreign origin. In response the animal produces antibodies, primarily immunoglobulin gamma, that are able to bind specifically to the antigen. Evidently a given antigen provides many different binding sites, thus giving rise to many different antibody molecules. If blood is removed from the animal and allowed to coagulate, antibodies can be isolated from the serum fraction. Even when purified by recombination with the original antigen, immunoglobulin gamma molecules produced in this way vary slightly.

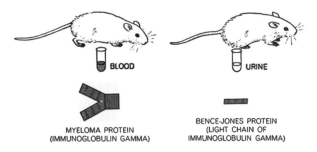

IDENTICAL ANTIBODY-LIKE MOLECULES are produced in large numbers by mice and humans who suffer from myelomatosis, a cancer of the cells that synthesize immunoglobulin. These abnormal immunoglobulins, all alike, can be isolated from the animal's blood (*left*). Often an abnormal protein also appears in the urine (*right*). Called a Bence-Jones protein, it seems to be the light chain of the abnormal immunoglobulin.

Immunoglobulin gamma has a molecular weight of about 150,000, corresponding to some 23,000 atoms, of which a carbohydrate fraction forms no more than 2 or 3 percent. Chemical studies have shown that the immunoglobulin gamma molecule is built up of four polypeptide chains, which, as in all proteins, are formed from strings of amino acids joined to one another through peptide bonds. The four chains are paired so that the molecule consists of two identical halves, each consisting of one long, or heavy, chain and one short, or light, chain. The four chains are held to one another by the disulfide bonds of the amino acid cystine. If the disulfide bonds are split, the heavy and light chains are still bound to each other. If, however, they are put in an acid solution or one containing a substance such as urea, they dissociate and can be separated by their difference in size.

Immunoglobulin gamma molecules can also be split by proteolytic enzymes such as papain, which breaks the molecule into three pieces of about equal size. Two, known as *Fab* (for "fragment antigen binding"), appear to be identical, and the third, known as Fc (fragment crystalline), is quite different. *Fab* is so named because it will still combine with the antigen although it will not precipitate with it. Each *Fab* fragment carries one combining site; thus the two fragments together account for the two combining sites that each antibody molecule had been deduced to possess. The Fc fragment prepared from rabbit immunoglobulin gamma crystallizes readily, but neither the *Fab* fragments nor the whole molecule has ever been crystallized.

Since crystals form easily only from identical molecules, it was guessed that the halves of the heavy chain that comprise the Fc fragment are probably the same in all molecules and that the complexity is mainly in the *Fab* fragments where the combining sites are found. The enzyme papain, which causes the split into three pieces, can hydrolyze a great variety of peptide bonds, and yet only a few in the middle of the heavy chain are in fact split; it looks as if in the *Fab* and Fc fragments the peptide chains are tightly coiled in such a way that the enzyme cannot gain access. This suggests a picture in which three compact parts of the molecule are joined by a short flexible section neat the middle of the heavy chain.

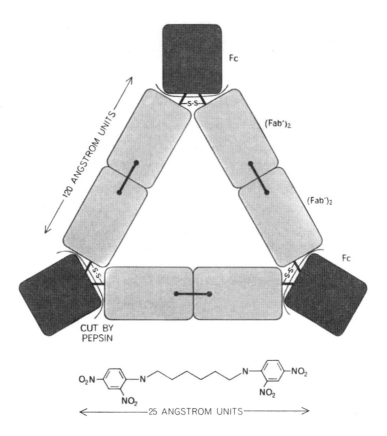

ANTIBODY-ANTIGEN COMPLEX seen in electron micrographs is thought to have this triangular structure. Below it, drawn to a large scale, is the synthetic antigen: a six-carbon chain carrying a dinitrophenyl group at each end. Three such antigen molecules appear able to bind together three immunoglobulin gamma molecules.

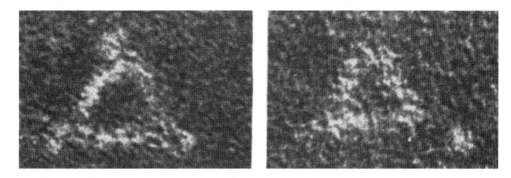

EFFECT OF PEPSIN COMPLEX is demonstrated in electron micrographs taken by Green and Valentine. In the normal complex formed by immunoglobulin gamma and the dinitrophenyl compound (*left*) a typical triangular structure contains a small lobe, or lump, at each corner, which is thought to be the Fc part of the immunoglobulin molecule. If the antibody is first treated with pepsin, which splits off the Fc fragment, the remaining (F*ab*')₂ molecule still reacts with the antigen but the corner lobes are missing (*right*).

The full structure of a protein molecule showing the arrangement in space of the peptide chains and the positioning of the amino acids along them can at present only be achieved by x-ray crystallography. Such work has been started at Johns Hopkins University with the Fc fragment. Electron microscopy, however, can provide much information about the shape of protein molecules, and successful electron microscope studies have been made recently with rabbit antibodies. When the antibodies are free, no clear pictures are obtained, which suggests that the molecules have a loose structure that is without definite shape. If they are combined with antigen, however, good pictures can be made. Michael Green and Robin Valentine of the National Institute for Medical Research in London prepared antibodies in rabbits that would combine with a benzene derivative known as a dinitrophenyl group. This can be done, as Karl Landsteiner showed many years ago, by injecting into the rabbit a protein on which dinitrophenyl groups have been substituted. Antibodies are formed, some of which combine specifically with the substituent dinitrophenyl coupled onto other proteins or into smaller molecules.

Green and Valentine investigated the smallest compound carrying two dinitrophenyl groups that would crosslink two or more antibody molecules. This proved to be a six-carbon chain with a dinitrophenyl group at each end. This material does not form a precipitate with antibody, but with the electron microscope one can see ringlike structures that appear to contain three to five antibody molecules. The small antigen molecule is not visible. The three-component structure is believed to consist of three antibody molecules linked by three molecules of invisible antigen. The lumps protruding from the corners are thought to be Fc fragments. This interpretation is supported by using the proteolytic enzyme pepsin to digest off the Fc fragment, leaving two Fab molecules held together by a disulfide bond and referred to as $(Fab')_2$. When these $(Fab')_2$ molecules are combined with antigen, rings are formed as before, but the lumps at the corners are now gone, confirming the idea that they were indeed the Fc part of the molecule.

Since most interest centers on the antibody combining site, the next problem to solve is whether the site is to be found in the light chain, which is entirely in the Fab fragment, or in the half of the heavy chain that is also present, or whether the site is formed by both chains together. It has not been possible to get a clear answer to this problem because the chains cannot be separated except in acid or urea solutions; this causes a partial loss of the affinity for antigen, which is not recovered even after the acid or urea is removed. Present evidence suggests that the heavy chain is the most important but that the light chain plays a role. This may be because it actually forms a part of the site or because it helps to stabilize the shape that the heavy chain assumes and hence plays a secondary role that may be only partially specific.

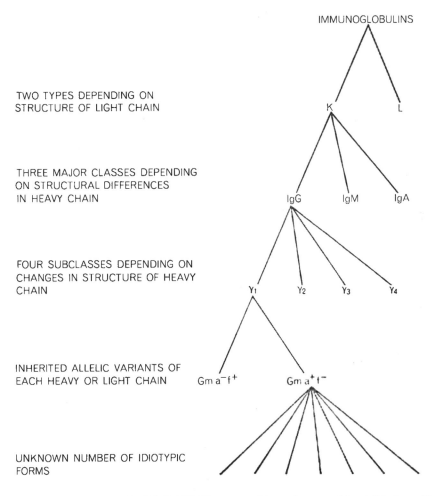

TWO TYPES DEPENDING ON
STRUCTURE OF LIGHT CHAIN

THREE MAJOR CLASSES DEPENDING
ON STRUCTURAL DIFFERENCES
IN HEAVY CHAIN

FOUR SUBCLASSES DEPENDING ON
CHANGES IN STRUCTURE OF HEAVY
CHAIN

INHERITED ALLELIC VARIANTS OF
EACH HEAVY OR LIGHT CHAIN

UNKNOWN NUMBER OF IDIOTYPIC
FORMS

**SUBDIVISIONS OF HUMAN IMMUNOGLOBULIN** presented investigators with a difficult problem to unravel. For simplicity, subdivisions are shown for only one branch at each level. The abbreviation *"IgG"* stands for immunoglobulin gamma, the antibody found in largest amounts and the one most easily isolated. Idiotypic forms are apparently unique to individual animals and may involve alterations in both the light and the heavy chains.

In any case, the field is clear for a direct attempt at comparative studies of the chemical structure of the light chain as well as of the half of the heavy chain that lies in the *Fab* part of the molecule. The shape and hence the specificity of the combining site must depend on the configuration of the peptide chains of the *Fab* fragment; this is believed to be determined only by the sequence of the different amino acids in the chain. Therefore it is reasonable to expect that if the amino acid sequence is worked out for the *Fab* half of the heavy chain and perhaps also for the light chain, then in some sections sequences will be found that determine the configuration of the combining site and that will be characteristic for each antibody specificity. Attempts to carry out such sequence studies, however, seemed unattractive because of convincing evidence that all preparations of immunoglobulin gamma—even samples of purified antibodies obtained by precipitation with a specific antigen—were actually mixtures of many slightly different molecules with presumably different amino acid sequences.

Although the complexity of immunoglobulin gamma (and of the other classes of immunoglobulins) has presented investigators with a most difficult puzzle, considerable progress has now been made in solving much of it. First, there are two kinds of light chain, named kappa and lambda, but in any one molecule both light chains are of the same type. The molecules containing kappa chains are known as *K* type and those with lambda chains as *L* type. Then in some species (probably in all) the immunoglobulin gamma class contains several subclasses; four have been identified in human gamma globulin. The subclasses differ in their heavy chains, which carry not only the characteristic features of the class but also small differences that distinguish the subclasses. In any one individual, molecules will be found of both *K* and *L* type, and they belong to all the subclasses. In addition each of the kinds of chain shows differences, known as allelic forms, that are inherited according to Mendelian principles. In an individual homozygous for this property only one allelic form of, say, the kappa chain will be present, but in a heterozygous individual there will be two forms of the kappa chain. It scarcely need be stressed that all these phenomena lead to a very complex mixture of molecules of immunoglobulin gamma in the serum of any individual. Yet there is still another kind of complexity termed idiotypic. In certain circumstances it is possible for an animal to synthesize antibody molecules that are unique to itself, distinct from other antibody molecules of the same specificity in other individuals of the same species—and distinct from all other immunoglobulins in its own blood.

Perhaps the most remarkable aspect of all of this is that the complexity seems to bear no relation to the structure of the antibody combining site. As far as we know at present, any antibody specificity may be found on any of these many different kinds of molecule.

All such variations are likely to be based on differences in amino acid sequence, and already some differences relating to subclass and allelic changes have been identified. The structural differences are so small, however, that it is not possible to separate out single kinds of molecule by the methods available for the fractionation of proteins.

Thus it was a great step forward when it was recognized that in certain forms of cancer, immunoglobulin molecules of apparently a single variety appear in the blood. Such immunoglobulins have only one type of light chain and one subclass of heavy chain, and each chain belongs to one or the other allelic form. As far as we know each chain has only one amino acid sequence and therefore belongs to only one idiotypic form.

The disease responsible for this unique production of antibody is known as myelomatosis. Observed in both mice and men, it is a cancer of the cells that synthesize immunoglobulin, often those in the bone marrow. Apparently a single cell, one of the great number that synthesize immunoglobulins, starts to divide rapidly and leads to an excessive production of a single kind of immunoglobulin. This provides evidence, incidentally, that the complexity of immunoglobulin molecules arises from their synthesis by many different kinds of cells. These abnormal immunoglobulins are known as myeloma proteins. Because they are often present in the blood in a concentration several times higher than all the other immunoglobulins together, they can be isolated rather easily.

Moreover, in about half of all myeloma patients an abnormal protein appears in the urine in large amounts. This substance was first observed by Henry Bence-Jones at Guy's Hospital in London in 1847 and has been known ever since as Bence-Jones protein. Its nature, however, was not recognized until five years ago, when Gerald M. Edelman and J. A. Gally of Rockefeller University and independently Frank W. Putnam of the University of Florida showed that Bence-Jones protein is probably identical with the light chains of the myeloma protein in the serum of the same patient. Because Bence-Jones proteins can be obtained easily, without any inconvenience to the patient, they were the first materials used for amino acid sequence studies.

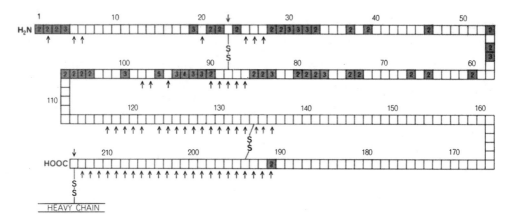

IMMUNOGLOBULIN LIGHT CHAIN, represented by analyses of human Bence-Jones proteins of the *K* type, has 214 amino acid units. Colored squares show where amino acids have been found to vary from one protein to another; blank squares show where no variation has yet been found. Numbers in the squares indicate how many different amino acids have been identified so far at a given site. Arrows mark positions where a particular amino acid has been found in at least five different proteins. Complete amino acid sequences are now known for three human Bence-Jones proteins and partial sequences for about 20 others. All variations occur in the first half of the chain with one exception, the variation at position 191. This is related to the allelic, or inherited, character of light *K* chains, hence differs from the alterations in the variable half of the chain. The diagram is based on one recently published by S. Cohen of Guy's Hospital Medical School in London and C. Milstein of the Laboratory of Molecular Biology in Cambridge.

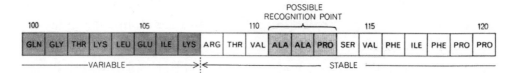

MIDDLE PART OF LIGHT CHAIN, as determined for one human Bence-Jones protein (*K* type), includes the amino acids at positions 111, 112 and 113 that are common to both *K*- and *L*-type Bence-Jones proteins of humans and to K-type Bence-Jones proteins of mice. It has been suggested that the section of the gene coding for this sequence may provide a special "recognition point" for the joining of two different genes responsible for the variable and stable sections of the light chain or, possibly, for bringing into play a mechanism to change the amino acid sequence in the variable section.

Although complete sequences have been worked out for only two Bence-Jones proteins in the mouse and only three human Bence-Jones proteins, perhaps 20 more have been partially analyzed. A remarkable fact has emerged. It seems that all Bence-Jones proteins of the same type have exactly the same sequence of amino acids in the half of the molecule that ends in the chemical group COOH (hence known as the C-terminal half) but show marked variation in the half that ends in the group $NH_2$ (the N-terminal half). Of 107 amino acid positions in this half, at least 40 have been found to vary. No two Bence-Jones proteins have yet been found to be identical in the N-terminal half, so that the possibility of molecular variation is clearly great. Given the possibility of variation at 40 sites and supposing that only two different amino acids can occupy these sites, it would be possible to construct $2^{40}$, or more than 10 billion, different sequences. Actually as many as five different amino acids have been found to occupy one of the variable sites.

The amino acid sequence studies of the heavy chain are less advanced than those with the Bence-Jones proteins because the material is more difficult to obtain and is more than twice the length. Results with the heavy chain of two human myeloma proteins, however, have shown them to have many differences in sequence for more than 100 amino acids from the N-terminal end, whereas the remainder of the chain appears to be identical in both cases. Accordingly it seems certain that the heavy chains will show the same phenomena as the light chains; it is possible that the length of the variable section in both chains will be similar.

Inasmuch as both variable sections are in the *Fab* fragment of the molecule it seems obvious that these sections must participate in creating the many different antibody combining sites. All the work discussed here has been done with myeloma proteins, and since each has a single amino acid sequence in both heavy and light chains, it would follow that each will be a specific antibody against one of an untold number of different antigenic sites. The chances, therefore, of finding a myeloma protein in which antibody specificity is directed to a known, well-defined antigenic site seemed small. Nevertheless, several myeloma proteins have recently been found to possess antibody-like activity against known antigens. A comparison of the sequences of their heavy and light chains may give a lead as to where the combining site is located.

I t has been believed with good reason that myeloma proteins are typical of normal molecules of immunoglobulin gamma, each being a homogeneous example of the many different forms present. It thus seemed likely that any attempt to determine the amino acid sequence of immunoglobulin gamma from a normal animal would be impossible, especially in the variable region that is of particular interest. One would expect normal immunoglobulin gamma to be a mixture of many thousands of different molecules, each with a different sequence in the variable region.

Amino acid sequences of polypeptide chains are found by using enzymes to break the chains into pieces from 10 to 20 amino acids long. It is then possible to work out the sequence of each piece. By using different enzymes the original chain can be broken at different places, with the result that some pieces overlap. This provides enough clues for the whole sequence

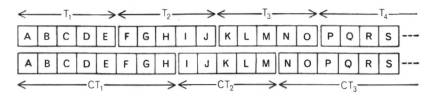

DETERMINATION OF AMINO ACID SEQUENCE in the polypeptide chains of proteins depends on the use of enzymes that cleave the chains into short fragments next to particular amino acids. The sequence in the resulting fragments can then be established. Thus trypsin might split a chain into fragments $T_1$, $T_2$, $T_3$ and $T_4$, Another enzyme, chymotrypsin, might split the same chain into fragments $CT_1$, $CT_2$ and $CT_3$, Since these fragments must overlap one can establish their order unequivocally and thereby the sequence of the entire chain.

IMMUNOGLOBULIN SEQUENCE should be amenable to analysis even though a particular antibody sample might contain a variety of slightly different molecules. Slight variations at certain positions should not prevent the ordering of similar fragments.

to be put together, rather like a one-dimensional jigsaw puzzle. When the protein is pure, there is only one order of amino acids possible, and all the sequences of the individual fragments will fit into it.

One can see that if this method were attempted with a protein that was in fact a mixture of many slightly different proteins, each with a different sequence, a hopelessly confusing picture would probably result. The work with the myeloma protein suggested, however, that there would be a constant part as well as a variable part, and it seemed worthwhile to see what progress could be made in determining at least the constant part. Work at Duke University showed that the whole of the Fc section of the heavy chain of normal rabbit immunoglobulin gamma gave a coherent sequence and was therefore part of the stable section, as had been expected. Recent work in our laboratory has now shown that the coherence continues well into the other half of the heavy chain. Although the work is far from complete, it seems possible that a full sequence will be established right through the entire heavy chain. Variations have been picked up in a number of positions and no doubt many more will be found, but the results are not completely confusing, as might have been expected if normal immunoglobulin gamma were a mixture of many thousands of myeloma proteins, each with substantially different sequences in the variable parts of the chain. The conflict between the results with the myeloma proteins and the recent results with normal rabbit immunoglobulin may be more apparent than real.

What does all this mean in terms of the structure of antibodies and their power to combine specifically with antigens? The phenomenon of a variable section and a stable section in both heavy and light chains is extraordinary and is unique to immunoglobulins; the variable section is in the part of the molecule known to contain the combining site. It therefore seems certain that this must be the basis of the specific configuration of the combining site.

It should be emphasized that all this work is very incomplete. In another year or so it will undoubtedly be much easier to see just how different one myeloma protein is from another in both the heavy and the light chains. It may be that the differences between any two will on the average be small, so that for a mixture of many molecules the amino acid in any one position will be common to 80 or 90 percent of the molecules. Presumably this explains how it is possible to find a comprehensible amino acid sequence in normal immunoglobulin gamma.

It may also be, however, that myeloma proteins are not quite typical of normal antibodies. Because they are the result of a disease they may exaggerate a normal phenomenon. Although they are invaluable in drawing attention to a fundamental mechanism, they may mislead us by exhibiting greater variability than is present in normal immunoglobulin gamma.

Whatever the answer, the existence of a stable section and a variable section, which has been shown so clearly in the Bence-Jones proteins and which also occurs in the heavy chains, is a remarkable phenomenon. The

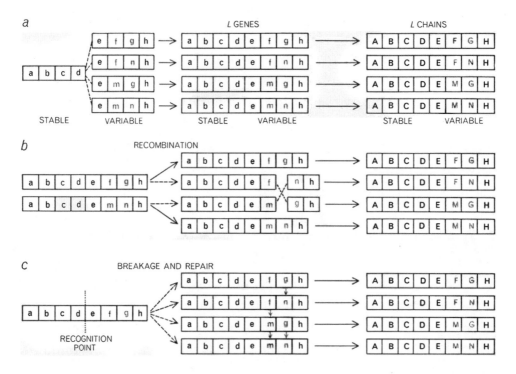

VARIABILITY OF IMMUNOGLOBULIN MOLECULES has been explained by three principal hypotheses. The simplest (*a*) suggests that one gene codes for the stable section of each chain and that a great number, perhaps hundreds of thousands, code for the variable section. A second idea (*b*) is that several genes are divided into stable and variable sections and that the latter inter change parts during cell division. A third proposal (*c*) suggests that there may he a recognition point in the gene and that an enzyme partially splits, or breaks, the gene on the variable side of that point. When repaired by other enzymes (*arrows*), mistakes are made, thus giving rise to many different amino acid sequences in the antibody molecule.

mechanism of its biological origin has aroused intense interest. Many hypotheses have been put forward, but there are perhaps three principal ones.

A straightforward mechanism would be to have a single gene coding for each stable section in the antibody molecule and as many genes as necessary (tens of thousands or hundreds of thousands) coding for the variable sections. The cell would also be provided with a means for fusing the product of the two kinds of gene to construct the complete immunoglobulin molecule. (In this as in the other suggestions, the presence of an antigen would somehow trigger production of the appropriate antibody.)

A second proposal invokes the concept of genetic recombination, which involves the exchange of parts of genes. One can imagine several genes that are divided into a stable portion and a variable portion. During cell division, when genes are pairing and duplicating, the variable portion would interchange sections, thereby giving rise to many different genes capable of coding the variable parts of the antibody molecule.

The third suggestion visualizes that the gene for, say, the light chain may contain a "recognition point" midway in its structure. This might provide a specific attachment site for an enzyme that can split the nucleic acid of the gene only on the side coding for the variable section. When the broken portion is repaired by other enzymes, mistakes are made, thereby giving rise to many different sequences of nucleotides—the nucleic acid building blocks that embody the genetic message. These differences are then translated into different amino acid sequences in the variable portion of the antibody molecule.

There is no clear answer as to which methods, if any or all, are the operative mechanisms, but a continuation of the structural studies may provide a clearer understanding. When this understanding is attained, it should lead to ideas about how to change, stimulate or suppress immune reactions as medical practice requires and therefore should be of great practical value as well as solving one of the most intriguing problems in biology.

## ABOUT THE AUTHOR

**R. R. Porter** (1917-1985) was awarded the Nobel Prize in Physiology or Medicine in 1972 along with Gerald M. Edelman for their discoveries concerning the chemical structure of antibodies. He put forward the peptide chain structure of antibodies in 1962.

# The Induction
# of Cancer by Viruses

*Normal cells cultured in glassware can be
transformed into cancer cells by several viruses.
Such "model" systems are studied to find how
the virus, with fewer than 10 genes,
can produce the change.*

## Renato Dulbecco

ancer, one of the major problems of modern medicine, is also a
fascinating biological problem. In biological terms it is the manifestation of changes in one of the more general properties of the
cells of higher organisms: their ability to adjust their growth rate to the
architectural requirements of the organism. To learn more about cancer is
therefore to learn more about this basic control mechanism. Over the past
decade dramatic advances in our knowledge of cancer have resulted from
the use of viruses to elicit the disease in simple model systems. A certain
understanding of the molecular aspects of cancer has been attained, and
the foundation has been laid for rapid progress in the foreseeable future.

A cancer arises from a single cell that undergoes permanent hereditary changes and consequently multiplies, giving rise to billions of similarly altered cells. The development of the cancer may require other conditions, such as failure of the immunological defenses of the organism. The fundamental event, however, is the alteration of that one initial cell.

There are two main changes in a cancer cell. One change can be defined as being of a regulatory nature. The multiplication of the cells of an animal is carefully regulated; multiplication takes place only when it is required, for example by the healing of a wound. The cancer cell, on the other hand, escapes the regulatory mechanisms of the body and is continuously in a multiplication cycle.

The other change of the cancer cell concerns its relations with neighboring cells in the body. Normal cells are confined to certain tissues, according to rules on which the body's overall architecture depends. The cancer cell is not confined to its original tissue but invades other tissues, where it proliferates.

The basic biological problem of cancer is to identify the molecular changes that occur in the initial cancer cell and determine what causes the changes. The particular site in the cell affected by the changes can be approximately inferred from the nature of the changes themselves. For example, a change in the regulation of cell growth and multiplication must arise from a change in the regulation of a basic process in the cell, such as the synthesis of the genetic material deoxyribonucleic acid (DNA). The alterations in relations with neighboring cells are likely to flow from changes in the outside surface of the cell, which normally recognizes and responds to its immediate environment.

Experimental work directed toward the solution of this central problem makes use of cancers induced artificially rather than cancers that occur spontaneously. Spontaneous cancers are not suitable for experiments because by definition their occurrence cannot be controlled; moreover, when a spontaneous cancer becomes observable, its cells have often undergone numerous changes in addition to the initial one. In recent years model systems for studying cancers have been developed by taking advantage of the fact that animal cells can easily be grown in vitro—in test tubes or boxes of glass or plastic filled with a suitable liquid medium. This is the technique of tissue culture.

Since the use of tissue culture has many obvious experimental advantages, methods for the induction of cancer in vitro have been developed. The most successful and most widely employed systems use viruses as the cancer-inducing agent. In these systems the initial cellular changes take place under controlled conditions and can be followed closely by using an array of technical tools: genetic, biochemical, physical and immunological.

It may seem strange that viruses, which are chemically complex structures, would be preferable for experimental work to simple cancer-inducing chemicals, of which many are available. The fact is that the action of cancer-inducing chemicals is difficult to elucidate; they have complex chemical effects on a large number of cell constituents. Furthermore, even if one were to make the simple and reasonable assumption that chemicals cause cancer by inducing mutations in the genetic material of the cells, the problem would remain enormously difficult. It would still be almost impossible to know which genes are affected, owing to the large number of genes in which the cancer-causing mutation could occur. It is estimated that there are millions of genes in an animal cell, and the function of most of them is unknown. With viruses the situation can be much simpler. As I shall show, cancer is induced by the genes of the virus, which, like the genes of animal cells, are embodied in the structure of DNA. Since the number of viral genes is small (probably fewer than 10 in the system discussed in this article), it should be possible to identify those responsible for cancer induction and to discover how they function in the infected cells. The problem can thus be reduced from one of cellular genetics to one of viral genetics. The reduction is of several orders of magnitude.

A number of different viruses have the 30 ability to change normal cells into cancer cells in vitro. In our work at the Salk Institute for Biological Studies we employ two small, DNA-containing viruses called the polyoma virus and simian virus 40 (SV 40), both of which induce cancer when they are inoculated into newborn rodents, particularly hamsters, rats and, in the case of the polyoma virus, mice. Together these viruses are referred to as the small papovaviruses.

In tissue-culture studies two types of host cell are employed with each virus. In one cell type-the "productive" host cell-the virus causes what is known as a productive infection: the virus multiplies unchecked within the cell and finally kills it. In another type of cell–the "transformable" host cell-the virus causes little or no productive infection but induces changes similar to those in cancer cells. This effect of infection is called transformation rather than cancer induction because operationally it is recognized

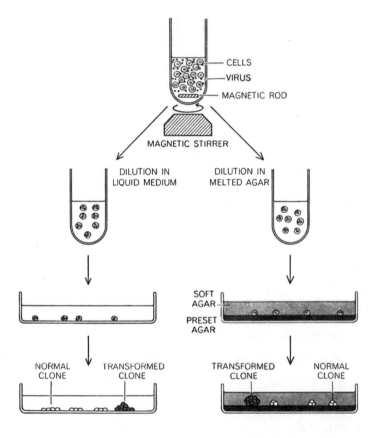

TRANSFORMATION EXPERIMENT produces cell colonies that differ in appearance, depending on the nature of the culture medium. BHK (hamster) cells are first incubated with polyoma virus for about an hour at 37 degrees centigrade, being stirred constantly. During this time viral particles enter the cells. The infected cells are then diluted either in a liquid medium or in melted agar and transferred to culture dishes. In the agar system the melted agar is poured on a layer of preset agar. The dishes are incubated at 37 degrees C. Cell colonies in the liquid medium develop in contact with the bottom of the container, whereas those in agar form spherical colonies above the preset agar layer. The results of using the two kinds of media are shown in the photographs on the opposite page.

from the altered morphology of the cells in vitro rather than from the production of cancer in an animal.

In the experimental work it is convenient to employ as host cells, particularly for transformation studies, permanent lines of cellular descent, known as clonal lines, that are derived from a single cell and are therefore uniform in composition. By using these clonal lines the changes caused by the virus can be studied without interference from other forms of cellular variation; one simply compares the transformed cells with their normal counterparts. Two lines that are widely employed are the "BHK" line, which was obtained from a hamster by Ian A. Macpherson and Michael G. P. Stoker of the Institute of Virology of the University of Glasgow, and the "3T3" line, which was obtained from a mouse by George J. Todaro and Howard Green of the New York University School of Medicine. BHK cells are particularly suitable for transformation by the polyoma virus; 3T3 cells are readily transformed by SV 40 and less easily by the polyoma virus.

In a typical transformation experiment a suspension of cells in a suitable liquid medium is mixed with the virus. The cells are incubated at 37 degrees centigrade for an hour; they are stirred constantly to prevent them from settling and clumping together. A sample of the cells is then distributed in a number of sterile dishes of glass or plastic that contain a suitable nutrient medium. The dishes are incubated at 37 degrees C. for a period ranging from one to three weeks. The cells placed in the dishes settle and adhere to the bottom. There they divide, each cell giving rise to a colony. If the number of cells is sufficiently small, the colonies remain distinct from one another and are recognizable to the unaided eye after about 10 days' incubation; they can be studied sooner with a low-power microscope. When the colonies are fully developed, they are usually fixed and stained. Colonies of normal and transformed cells can be recognized on the basis of morphological characters I shall describe. By picking a colony of transformed cells and reseeding its cells in a fresh culture, clonal lines of transformed cells can be easily prepared. The transformation of BHK cells can also be studied by a selective method that involves suspending the cells in melted agar, which then sets. Transformed cells give rise to spherical colonies, visible to the unaided eye, whereas normal cells grow little or not at all.

C olonies of transformed cells, and cultures derived from such colonies, differ morphologically from their normal counterparts in two obvious ways; these differences show that changes have occurred in the regulatory properties of the cells and also in the way they relate to their neighbors. The transformed cultures are thicker because they continue to grow rapidly, whereas normal cultures slow down or stop; in addition the transformed cells are not regularly oriented with respect to each other because they fail to respond to cell-to-cell contact. The altered response to contacts can be best appreciated in time-lapse motion pictures of living cultures.

In sparse BHK cultures the cells move around actively; if a cell meets another cell in its path, it usually stops moving and slowly arranges itself in contact with and parallel to the other cell. In this way a characteristic pattern of parallel lines and whorls is generated, since the cells do not climb over each other. In a culture of a derivative of the BHK line transformed by polyoma virus the same active movement of the cells is observed. When a cell meets another in its path, however, it continues to move, climbing over the other. In this way the arrangement of the cells becomes chaotic, without any discernible pattern.

These alterations of the transformed cells indicate their intimate relatedness to cancer cells. The relatedness is shown in a more dramatic way by the ability of the transformed cells to grow into a cancer when injected, in sufficient number, into a live host that does not present an insurmountable immunological barrier to their survival. For example, BHK cells, which were originally obtained from a hamster, can be transplanted into hamsters; similarly, cells of inbred strains of mice can easily be transplanted into mice of the same strain. The injection of roughly a million transformed cells into a hamster or mouse will be followed by the development of a walnut-sized tumor at the site of inoculation in about three weeks. Untransformed cells, on the other hand, fail to produce tumors.

A crucial finding is that the transformation of healthy cells is attributable to the genes present in the viral DNA that penetrates the cells at infection. The viral genes are the units of information that determine the consequences of infection, Each viral particle contains a long, threadlike molecule of DNA wrapped in a protein coat. Each of these molecules is made up of two strands twisted around each other. Attached to the molecular backbone of each strand is the sequence of nitrogenous bases that contains the genetic information of the virus in coded form. There are four kinds of base, and the DNA molecule of a papovavirus has some 5,000 bases on each strand. Each species of virus has a unique sequence of bases in its DNA; all members of a species have the same base sequence, except for isolated differences caused by mutations.

The double-strand molecule of DNA is so constructed that a given base in one strand always pairs with a particular base in the other strand; these two associated bases are called complementary. Thus the two DNA strands are also complementary in base sequence. Complementary bases form bonds with each other; the bonds hold the two strands firmly together. The two strands fall apart if a solution of DNA is heated to a fairly high temperature, a process called denaturation. If the heated solution is then slowly cooled, a process called annealing, the complementary strands unite again and form doublestrand molecules identical with the original ones.

When suitable cells are exposed to a virus, a large number of viral particles are taken up into the cells in many small vesicles, or sacs, which then accumulate around the nucleus of the cell. Most of the viral particles remain inert, but the protein coat of some is removed and their naked DNA enters the inner compartments of the cell, ultimately reaching the nucleus. Evidence that cell transformation is caused by the viral DNA, and by the genes it carries, is supplied by two experimental results.

The first result is that cell transformation can be produced by purified viral DNA, obtained by removing the protein coat from viral particles; this was first shown by G. P. Di Mayorca and his colleagues at the Sloan-Kettering Institute. The extraction of the DNA is usually accomplished by shaking the virus in concentrated phenol. In contrast, the empty viral coats do not cause transformation. These DNA-less particles are available for experimentation because they are synthesized in productively infected cells together with the regular DNA-containing particles. The empty coats have a lower density than the complete viral particles; hence the two can be separated if they are spun at high speed in a heavy salt solution, the technique known as density-gradient centrifugation.

A more sophisticated experiment performed at the University of Glasgow also rules out the possibility that the transforming activity resides in contaminant molecules present in the extracted DNA. The basis for this experiment is the shape of the DNA molecules of papovaviruses. The ends of each molecule are joined together to form a ring. When the double-strand filaments that constitute these ring molecules are in solution, they form densely packed supercoils. If one of the strands should suffer a single break, the supercoil disappears and the molecule becomes a stretched ring. Supercoiled molecules, because of their compactness, settle faster than

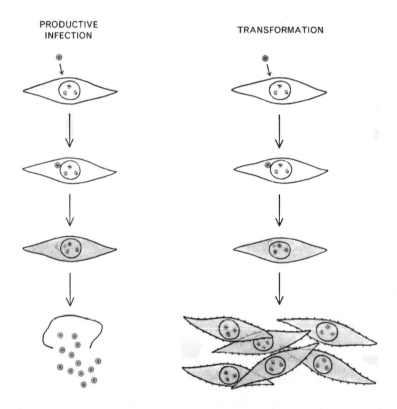

PRODUCTIVE INFECTION

TRANSFORMATION

VIRAL INVASION OF CELLS can have two different results. One result is "productive infection" (*left*), in which viral particles mobilize the machinery of the cell for making new viral particles, complete with protein coats. The cell eventually dies, releasing the particles. The other result is transformation (*right*), in which the virus alters the cell so that it reproduces without restraint and does not respond to the presence of neighboring cells. Viral particles cannot be found in the transformed cells. The tint of color in these cells indicates the presence of new functions induced by the genes of the virus. The change in the cell membrane (*lower right*) denotes the presence of a virus-specific antigen.

stretched-ring molecules when they are centrifuged. Thus the two molecular types can be separated in two distinct bands.

By this technique polyoma virus DNA containing both molecular types can be separated into fractions, each of which contains just one type. Examination of the biological properties of these fractions shows that the transforming efficiency is strictly limited to the two bands of the viral DNA. Similarly, only the material in the two bands will give rise to productive infection. This result, among others, rules out the possibility that transformation is due to fragments of cellular DNA, which are known to be present in some particles of polyoma virus and therefore contaminate the preparations of viral DNA. The contaminant molecules have a very different distribution in the gradient.

The second result demonstrates directly that the function of a viral gene is required for transformation, by showing that a mutation in the viral genetic material can abolish the ability of the virus to transform. This important finding was made by Mike Fried of the California Institute of Technology, who studied a temperature-sensitive mutant line of polyoma virus called Ts-a. The virus of this line behaves like normal virus in cells at 31 degrees C., causing either transformation or productive infection, depending on the cells it infects. At 39 degrees C., however, the effect of the mutation is manifest, and the virus is unable to cause either transformation or infection; it is simply inactive.

We can now inquire whether the viral gene functions needed to effect transformation are transient or continuous. In other words, do the genes act only once and produce a permanent transformation of the cell line or must they act continuously to keep the cell and its descendants transformed?

A result pertinent to this question is that the transformed cells contain functional viral genes many cell generations after transformation has occurred, although they never contain, or spontaneously produce, infectious virus. The presence of viral genes has been demonstrated particularly well by T. L. Benjamin at Cal Tech, who has shown that the transformed cells contain virus-specific ribonucleic acid (RNA). To make the significance of this finding clear it should be mentioned that the instructions contained in the base sequence of the DNA in cells or viruses are executed by first making a strand of RNA with a base sequence complementary to that of one of the DNA strands. This RNA, called messenger RNA, carries the information of the gene to the cellular sites where the proteins specified by the genetic information are synthesized. Each gene gives rise to its own specific messenger RNA. If one could show that viral messenger RNA were present in transformed cells, one would have evidence not only that the cells contain viral genes but also that these genes are active. The viral RNA molecules can be recognized among those extracted from transformed cells, which are mostly cellular RNA, by adding to the mixture of RNA molecules heat-denatured, single-strand viral DNA. When the mixture of RNA and DNA is annealed, only the viral molecules of RNA enter into double-strand molecules with the viral DNA. The reaction is extremely sensitive and specific.

It is likely, therefore, that the viral genes persisting in the transformed cells are instrumental in maintaining the transformed state of the cells. This idea is supported by the observation that the form of the transformed

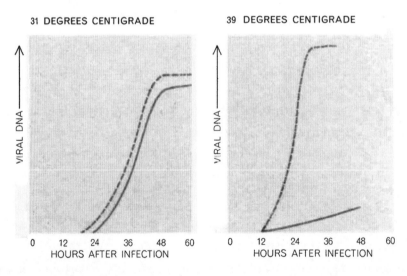

TEMPERATURE-DEPENDENT STRAINS OF POLYOMA VIRUS act normally at a temperature of 31 degrees C. (*left*) but exhibit mutated behavior at 39 degrees C. (*right*). The solid curves show the amount of viral deoxyribonucleic acid (DNA) synthesized in productive host cells containing the mutant virus. Broken curves show the viral DNA output in cells containing "wild type" (ordinary) polyoma virus. The mutant virus is called Ts-a.

cells is controlled by the transforming virus. This is seen clearly in cells of the line 3T3, which can be transformed by either SV 40 or polyoma virus. The transformed cells, although descended from the same clonal cell line, are strikingly different. Similar differences are also observed in other cell types transformed by the two viruses. Since the cells were identical before infection, the differences that accompany transformation by two different viruses can be most simply explained as the result of the continuing function of the different viral genes in the same type of cell. In fact, it is difficult to think of a satisfactory alternative hypothesis.

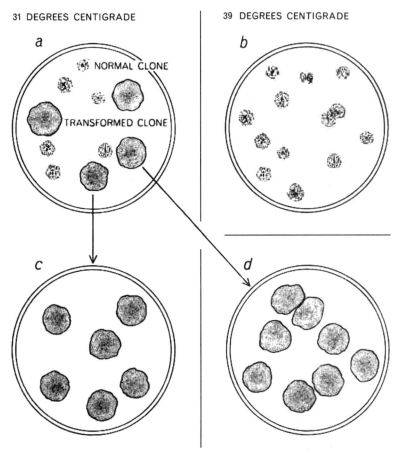

TRANSIENT ROLE OF TS-A GENE, which gives rise to temperature-dependent mutants of polyoma virus, can be demonstrated by raising the temperature of experimental cultures after the cells have been transformed by the virus. The mutant virus is able to transform cells at low temperature (*a*) but not at high temperature (*b*). Transformed cell colonies, or clones, remain transformed, as expected, at low temperature (*c*), but they also remain transformed when the temperature is raised (*d*). This experiment provides evidence that the Ts-a gene is needed for the initial transformation of the cell but is not needed thereafter.

It must be clear, however, that there is no conclusive evidence for this continuing role of the viral genes. It is therefore impossible to exclude an entirely different interpretation of the observation. One can argue, for example, that the persistence of the viral genes is irrelevant for transformation, and that the genes remain in the cells as an accidental result of the previous exposure of the cells, or of their ancestors, to the virus. Indeed, under many other circumstances viruses are often found in association with cells without noticeably affecting them. A conclusive clarification of the role of the persisting viral genes is being sought by using temperature-dependent viral mutants analogous to the Ts-a mutant I have mentioned. A virus bearing a temperature-dependent mutation in a gene whose function is required for maintaining the cells in the transformed state would cause transformation at low temperature. The cells, however, would revert to normality if the ambient temperature were raised. A small-scale search for mutants with these properties has already been carried out in our laboratory but without success; a large-scale search is being planned in several laboratories.

It should be remarked that no protein of the outer coat of the viral particles is ever found in the transformed cells. Thus the gene responsible for the coat protein is always nonfunctional. This could be either because the transformed cells have an incomplete set of viral genes and the coat gene is absent or because some genes remain "silent." The silence of these genes in turn could be attributed to failure either of transcription of the DNA of the gene into messenger RNA or of translation of the messenger RNA into protein. If failure of transcription were the mechanism, transformed cells would be similar to lysogenic bacteria. Such bacteria have a complete set of genes of a bacteriophage (a virus that infects bacteria), but most of the viral genes are not transcribed into RNA. No other significant similarities exist, however, between ordinary lysogenic bacteria and transformed cells; therefore it is more likely that the coat gene is either absent or, if it is present, produces messenger RNA that is not translated into coat protein. Whatever the mechanism, the lack of expression of the coat-protein gene, and probably of other genes as well, is essential for the survival of the transformed cells, since it prevents productive infection that would otherwise kill the cells.

S o far we have considered the genes of the virus in abstract terms. Let us now consider them in concrete ones by asking how many genes each viral DNA molecule possesses and what their functions are. The function of a viral gene is the specification, through its particular messenger RNA, of a polypeptide chain, which by folding generates a protein subunit; the subunits associate to form a functional viral protein. The final product can be an enzyme or a regulator molecule that can control the function of other genes (viral or cellular), or it can be a structural protein such as the coat protein of the viral particles.

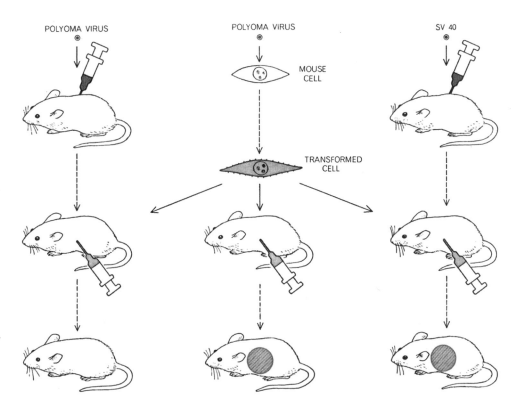

IMMUNIZATION EXPERIMENT shows that an animal will not develop a tumor after receiving a massive injection of transformed cells if it has previously received a mild inoculation of the virus used to transform the cells. Thus the animal at left, which has been immunized by an injection of polyoma virus, does not develop a tumor when injected with cells transformed by polyoma virus. The animal in the middle, not so immunized, develops a tumor following the injection of polyoma-transformed cells. The animal at right, which has received an injection of a different virus, SV 40, is not immunized against cells transformed by polyoma virus, hence it too develops a tumor. It would not develop a tumor, however, if injected with cells transformed by SV 40. Cells transformed by either polyoma virus or SV 40 contain a new antigen in their surface that makes them foreign to the animal strain from which they derive and therefore subject to its immunological defenses. These defenses can be mobilized by direct injection of the virus.

As I have said, each strand of the DNA of the small papovaviruses contains about 5,000 bases. Three bases are required to specify one amino acid, or one building block, in a polypeptide chain; therefore 5,000 bases can specify some 1,700 amino acids. It can be calculated from the total molecular weight of the coat protein of the viral particles and from the number of subunits it has that between a third and a fourth of the genetic information of the virus is tied up in specifying the coat protein. This genetic information is irrelevant for transformation, because no coat protein is made in the transformed cells. What remains, therefore, is enough genetic information to specify about 1,200 amino acids, which can constitute from four to eight small protein molecules, depending on their size. This is the

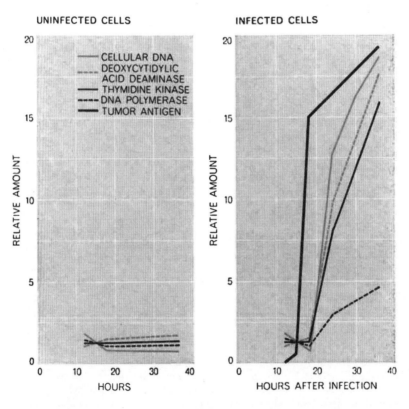

ACTIVATION OF DNA SYNTHESIS, along with activation of enzymes needed for its production, is a major consequence of viral infection of animal cells. Resting, uninfected cells make little DNA or enzymes associated with its synthesis (left). The values plotted are for kidney cells of an African monkey. When the cells are infected with SV 40, the output of DNA and associated enzymes rises steeply (right). Before these cellular syntheses are activated a new virus-specific protein, the "T antigen," whose role is unknown, appears.

maximum number of viral genetic functions that can be involved in the transformation of a host cell.

In order to discover these viral genetic functions the properties of normal cells have been carefully compared with the properties of cells that have been either transformed or productively infected. Characteristics present in the infected cultures can be considered to result, directly or indirectly, from the action of viral genes. We shall call these new characteristics "new functions." In this way six new functions have been discovered in the infected cells, in addition to the specification of the viral coat protein. Some of the new functions can be recognized biochemically; others can be shown by immunological tests to act as new cellular antigens.

The genetic studies with the papovaviruses have not gone far enough to reveal whether each of the new functions indeed represents the function of a separate viral gene, or whether all the gene functions have been identified. On the basis of the possible number of genes and the number of new functions it is likely that most, if not all, of the gene functions have been detected. At present the new functions are being attributed to the genes, and a large-scale effort is being made to produce temperature-sensitive mutants that will affect each of the genes separately. By studying the effect of such mutations on transformation it will be possible to establish the role of each gene in an unambiguous way.

For the moment we must limit ourselves to examining the various new functions and making educated guesses about their possible role in transformation. If transformation is continuously maintained by the function of viral genes, two new functions are particularly suspect as agents of transformation. One function involves a virus-specific antigen present on the surface of transformed cells; the other is the activation of the synthesis of cellular DNA and of cellular enzymes required for the manufacture of DNA by productively infected cells.

The induction of a virus-specific antigen on the cell surface was detected independently by Hans Olof Sjögren of the Royal Caroline Institute in Stockholm and by Karl Habel of the National Institutes of Health. They have shown that if an animal is inoculated with a mild dose of SV 40 or polyoma virus, it will develop an immune response that will enable it to reject cells transformed by the virus. Whereas the cells grow to form a tumor in the untreated animals, they are immunologically rejected by and form no tumor in the immunized animals. Rejection occurs only if the animals were immunized by the same virus used for transforming the cells. For instance, immunity against cells transformed by polyoma virus is induced by polyoma virus but not by SV 40, and vice versa. This shows that the antigen is virus-specific. The antigenic change is an indication of structural changes in the cellular surface, which may be responsible for the altered relations of transformed cells and their neighbors.

The activation of cellular syntheses, discovered independently in several laboratories, can be demonstrated in crowded cultures. If the cells in the culture are uninfected, they tend to remain in a resting stage. In these cells the synthesis of DNA, and of enzymes whose operation is required for DNA synthesis (such as deoxycytidylic acid deaminase, DNA polymerase and thymidine kinase), proceeds at a much lower rate than it does in growing cells. After infection by a small papovavirus a burst of new synthesis of both DNA and enzymes occurs; a viral function thus activates a group of cellular genes that were previously inactive. If the infection of the cells is productive, the activation of cellular syntheses occurs before the cells are killed. The activating viral function must act centrally, presumably at the level of transcription or translation of cellular genes that receive regulatory signals from the periphery of the cell; the signals themselves should be unchanged, since the cell's environment, in which the signals originate, is not changed. If the viral gene responsible for the activating function persists and operates in the transformed cells, it will make the cells insensitive to regulation of growth. Direct evidence for the operation of this mechanism in the transformed cells, however, has not yet been obtained.

A third viral function may be connected with such activation. This is the synthesis of a protein detected as a virus-specific antigen and called the T antigen (for tumor antigen). This antigen, discovered by Robert J. Huebner and his colleagues at the National Institute of Allergy and Infectious Diseases, differs in immunological specificity from either the protein of the viral coat or the transplantation antigen. The T antigen is present in the nucleus of both productively infected and transformed cells. In productive infection the T antigen appears before the induction of the cellular syntheses begins, and before the viral DNA replicates. Therefore the T antigen may represent a protein with a control function; for instance, it may be the agent that activates the cellular syntheses. For this assumption also direct evidence is lacking.

A fourth viral function relevant to transformation is the function of the gene bearing the Ts-a mutation, which we can call the Ts-a gene. The reader will recall that a virus line carrying this mutation transforms cells at low temperature but not at high temperature. Cells transformed at low temperature, however, remain transformed when they are subjected to the higher temperature, in spite of the inactivation of the gene. Thus the function of the Ts-a gene is only transiently required for transformation. In order to evaluate the significance of this result we must also recall that in productive infection the function of the Ts-a gene is required for the synthesis of the viral DNA. Therefore the transient requirement of this function in transformation may simply mean that the viral DNA must replicate before transformation takes place. If so, the Ts-a gene is not directly involved in transformation.

Another interpretation is possible. The function of the Ts-a gene is likely to be the specification of an enzyme involved in the replication of the viral DNA, for example a DNA polymerase, or a nuclease able to break the viral DNA at specific points, or even an enzyme with both properties. The action of a specific nuclease seems to be required for the replication of the viral DNA because the viral DNA molecules are in the form of closed rings. A nuclease, in breaking one of the strands, could provide a swivel around which the remainder of the molecule could rotate freely, allowing the two strands to unwind. The enzyme, although required for the replication of the viral DNA, may also affect the cellular DNA, for instance by causing breaks and consequently mutations. Such breaks have been observed in the DNA of cells that have been either productively infected or transformed by papovaviruses. If the Ts-a gene indeed acts on the DNA of the host cell, it could play a direct role in the transformation of the cell. Its actions would appear to be transient, however, since mutations in the cellular DNA would not be undone if the Ts-a gene were subsequently inactivated by raising the temperature of the system. A more definite interpretation of the Ts-a results must await the completion of the biochemical and genetic studies now in progress in several laboratories.

The last two of the six new viral functions observed in infected cells are not sufficiently well known to permit evaluation of their possible roles

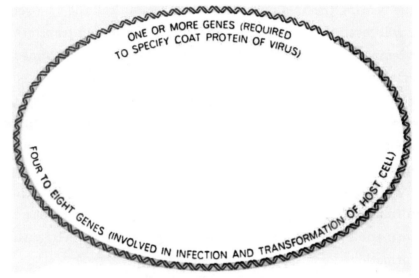

ONE OR MORE GENES (REQUIRED TO SPECIFY COAT PROTEIN OF VIRUS)

FOUR TO EIGHT GENES (INVOLVED IN INFECTION AND TRANSFORMATION OF HOST CELL)

**SEVEN FUNCTIONS IDENTIFIED WITH VIRUS ACTIVITY**

1. Specification of antigen found on surface of transformed cells

2. Specification of factor that activates synthesis of cellular DNA

3. Specification of antigen (T antigen) found in nuclei of infected and transformed cells.

4. Specification of enzyme involved in initial replication of viral DNA. (Attributed to the Ts-a gene.)

5. (Facilitation of cell infection by other viruses.)

6. (Induction of thymidine kinase enzyme.)

7. (Specification of coat protein of virus.)

SEVEN VIRAL FUNCTIONS have been identified in the infection and transformation of cells. The DNA present in the polyoma virus and SV 40 takes the form of a single ring-shaped molecule consisting of two helically intertwined strands (*top*). Each strand contains some 5,000 molecular subunits called bases that embody the genetic information of the virus in coded form. These bases, in groups of three, specify the amino acids that link together to form protein molecules. Thus 5,000 bases can specify some 1,700 amino acids, or enough to construct some six to 12 proteins. By definition it takes one gene to specify one protein. It is estimated that a third to a fourth of the bases in the viral DNA are needed to specify the protein in the coat of the virus. The remaining bases, enough for four to eight genes, specify the proteins involved in infection and transformation. Little is yet known about the fifth function in this list of seven. Functions 6 and 7 are not involved in cell transformation.

in cell transformation. One of these two functions is the induction of a thymidine kinase enzyme that is different from the enzyme of the same type normally made by the host cell. Thymidine kinase participates at an early stage in a synthetic pathway leading to the production of a building block required in DNA synthesis. There are reasons to believe, however, that the thymidine kinase induced by the virus may have a general regulatory effect in activating the DNA-synthesizing machinery of the cell after infection. One reason is that the viral thymidine kinase has not been found in transformed cells. Since this enzyme is induced by many viruses containing DNA, whether or not they cause transformation, its induction by the papovaviruses may be connected exclusively with productive infection.

The last new function is one observed so far only with SV 40. After cells have been productively infected with this virus they are changed in some way so that they become productive hosts for a completely different kind of virus, an adenovirus, even though they are normally not a suitable host for such viruses. Little is known about the biochemical steps involved.

The central mechanism of cell trans-formation and cancer induction would appear to be contained within the half-dozen viral functions I have discussed, perhaps together with a few others as yet unknown. Thus the problem is narrowly restricted. It is likely that the dubious points still remaining will be resolved in the near future, since the dramatic advances of the past several years have set the stage for rapid further progress.

This article should not be concluded without an attempt's being made to answer a question that will undoubtedly have arisen in the minds of many readers: Why are viruses able to induce cancer at all? For the two viruses discussed in this article, at least, it seems likely that the viral functions that are probably responsible for cell transformation have been selected by evolutionary processes to further the multiplication of the virus. Because the virus is small and cannot contain much genetic information it must exploit the synthetic mechanisms of the cell, including a large number of cellular enzymes, to achieve its own replication. Furthermore, in the animal hosts in which these viruses normally multiply, most cells that can undergo productive infection are in a resting stage and have their DNA-synthesizing machinery turned off. Thus the evolution of a viral function capable of switching on this machinery is obviously quite advantageous to the virus. This function must be very similar to the function of the cellular gene that regulates cellular DNA synthesis (and overall growth) in the absence of viral infection. The functions of the viral gene and of the cellular gene, however, must differ in one point, again for selective reasons: the cellular function must be subject to control by external signals, whereas the viral function must not be. The virus-induced alteration of the cellular surface seems also to be connected, in a way not yet understood, with viral multiplication, since in many viral infections viral proteins appear on the surface of cells.

The cancer-producing action of the papovaviruses can therefore be considered a by-product of viral functions developed for the requirements of viral multiplication. These viral functions lead to cancer development because they are similar to cellular functions that control cell multiplication, but they somehow escape the regulatory mechanisms that normally operate within the cell.

## ABOUT THE AUTHOR

**Renato Dulbecco** (born 1914) was awarded the Nobel Prize in Physiology or Medicine in 1975 along with David Baltimore and Howard M. Temin for their discoveries concerning the interaction between tumour viruses and the genetic material of the cell.

# The Structure and Function of Antibodies

*The complete amino acid sequence
of an immunoglobulin molecule has been
determined, defining the structure of antibodies
and providing information on their evolution
and differentiation and how they work.*

## Gerald M. Edelman

**I**mmunity" is an everyday word, ordinarily applied to the elaborate set of responses by which the body defends itself against invading microorganisms or foreign tissues. There is much more to immunity than its clinical aspects, however. What we have come to know about the immune system and its key molecules, antibodies, makes it apparent that immunology bears directly on some very fundamental problems: the nature of the mechanisms whereby molecules recognize one another, the manner in which genes are expressed in higher organisms and the origin of a variety of disease states, including cancer. In one way or another the solution of these deep problems will require an understanding of the structure of antibody molecules.

Antibodies have been known since the classic studies of Emil von Behring in the late 19th century. Only recently, however, have we begun to understand how vertebrate organisms recognize the sometimes subtle chemical differences between their own molecules and foreign molecules, which are termed antigens. Our insights rest on three serial developments. The first was the demonstration by Karl Landsteiner in 1917 that animals could form antibodies against certain small organic chemicals of known structure. By manipulating and modifying these "haptens," Landsteiner and others showed that antibodies distinguish among different antigens by recognizing differences in their shape. Moreover, the early studies implied that the number of different antibodies any single animal can make must be very large indeed. If an animal could make antibodies that could specifically bind a synthetic hapten the animal or its ancestors had never encountered before in nature, it seemed likely that the animal could make antibodies to almost any foreign antigen. Further work has largely supported this inference: most vertebrates are indeed capable of making many thousands of different antibodies. The second fundamental development was the bold idea of selective immunity advanced in the late 1950's by Niels K. Jerne and Sir Macfarlane Burnet, who proposed that the body already has all the information for making any of its antibodies *before* it ever encounters an antigen. The third development was the analysis of the structure of antibodies, which has taken place largely in the past decade. Before discussing this last development in detail, it will be useful to outline the seminal idea of selective immunity and some of the advances made by cellular immunologists.

The main notion, as implied above, is that the cells of the antibody-forming system have among them all the information they need to make any antibody molecule before they ever encounter any antigen. The antigen molecule does not instruct antibody-producing cells to shape the antibody molecule to fit it. Instead it selects cells that are already making antibodies that happen to fit. Then it stimulates those cells to make large quantities of the antibodies.

Antibodies can be likened to ready-made suits. The antigen is a buyer who decides to pick a number of different suits that fit more or less well rather than instruct a tailor to make one suit to fit him to order. To be well satisfied, the buyer must patronize a store with a very large stock of suits in a great variety of sizes and styles. The immune system is like a store with an almost unlimited stock, one ready to please any possible customer. This analogy fails in one important respect: to be complete it should provide that after each somewhat different ready-made suit is picked the manufacturer would proceed to make thousands of exact copies of it.

In a simplified picture of the cellular mechanisms corresponding to this selective response, each cell makes only one kind of antibody, which has an antigen-binding site of a particular shape. Presumably that antibody is located at the cell surface. An antigen injected into the body "tries on" different shapes. If a particular antibody "fits" more or less well, the cell making it divides and matures. Its progeny then make many more copies of the identical kinds of antibody, which may then be released into the blood to carry out their function of defense or of tissue rejection. Notice that this is a form of molecular recognition machine and that the specificity of recognition rests both on the presentation of a variety of antibodies and on the capacity of the cells to "amplify" the results of a recognition event. How are these requirements accomplished at a molecular level? In order to answer that question satisfactorily we must know a good deal about the structure of the antibody molecule itself.

A decade ago a number of experimental observations began to shed light on the details of antibody structure. The advance came when the results of sophisticated protein chemistry and genetic analysis fitted together with a classical observation about the products of a certain cancer. At the Rockefeller Institute in 1959 I found that antibody molecules consisted of polypeptide chains, or protein subunits, of more than one kind, and that the chains could be separated from one another by chemical means and studied in detail. These chains were called light chains and heavy chains because of the difference in their size, or molecular weight. At the same time R. R. Porter, now at the University of Oxford, showed that the antibody molecule could be cut into three different pieces by enzymes that cleave polypeptide chains. Two of them, termed "fragment antigen binding" (F*ab*), were identical and would still combine with antigen. The third, "fragment crystalline" (F*c*), was quite different: it would not combine with antigen but could be crystallized readily [see "The Structure of Antibodies," by R. R. Porter; SCIENTIFIC AMERICAN, October, 1967]. The observations on the two polypeptide chains and on the two fragments were the starting

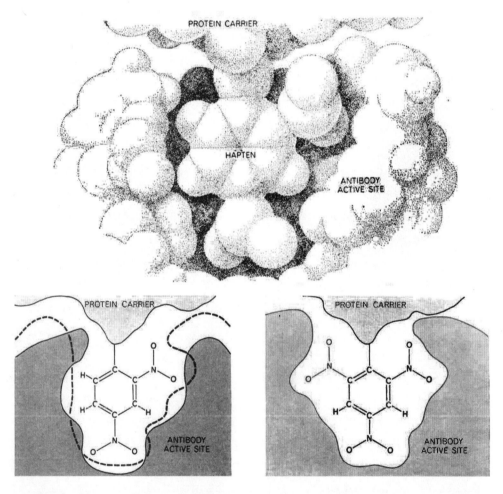

ANTIBODY recognizes a specific antigen by its fit. The top drawing shows the fit between the combining site on an antibody and a "hapten" antigen, a dinitrophenyl group, on a protein carrier. The bottom drawings show that two different antibody contours (*gray shape and black line*) can fit the dinitrophenyl group, one better than the other (*left*). If a third nitro group were on the hapten (*right*), those two antibodies would not fit; a third antibody with a different antigen-combining site would fit this picryl antigen. Because of picryl's similarity to dinitrophenyl, the third antibody would also fit the original hapten, but less precisely.

point for a series of investigations in many laboratories into the details of antibody structure.

Antibodies, it was known, were a family of proteins with a number of properties in common, found in the gamma globulin fraction of the blood. Unlike all proteins whose structure had been determined, antibody molecules were known to be very "heterogeneous": no single sequence of amino acids—the building blocks of proteins—could represent the polypeptide chains of antibodies, as can be done, for example, in the case of a "homogeneous" protein such as insulin or hemoglobin. This fact, together with the large size of antibody molecules, made it infeasible to carry out detailed chemical analyses of such molecules or to determine exactly how antibodies that bound to different antigens differed from one another. Fortunately the structural studies of the polypeptide chains of the immunoglobulin molecule led to a clue that made it possible to bypass the problem raised by the intrinsic heterogeneity of antibodies. The clue had to do with the nature of certain homogeneous proteins made by tumors of plasma cells, the cells that ordinarily produce the most antibodies.

Knowledge of these tumors goes back to 1847, when Henry Bence Jones, physician at St. George's Hospital in London, published a paper titled "On a new substance occurring in the Urine of a patient with Mollities Ossium." It began as follows: "On the 1st of November 1845 I received from Dr. Watson the following note, with a test tube containing a thick, yellow, semi-solid substance:—'The tube contains urine of very high specific gravity; when boiled it becomes highly opake; on the addition of nitric acid it effervesces, assumes a reddish hue, becomes quite clear, but, as it cools, assumes the consistence and appearance which you see: heat reliquifies it. What is it?'"

Jones verified the peculiar thermosolubility properties of the protein, subjected it to a careful elementary analysis and concluded that it was the "hydrated deutoxide of albumin."

In succeeding years many attempts were made to answer Dr. Watson's question in a definitive way, but although some 700 papers on the subject appeared in the ensuing century, Bence Jones proteins remained a kind of medical and biochemical curiosity except in the domain of practical diagnosis: the demonstration of the protein in the urine called for a diagnosis of multiple myeloma, the malignant disease of plasma cells that was formerly called mollities ossium. This disease is usually associated with malignant proliferation of plasma cells, excessive production of serum gamma globulins called myeloma proteins, bone lesions, disturbances of calcium metabolism, kidney disorders and often, of course, excretion of the characteristic Bence Jopes proteins.

It seemed that these proteins had something to do with immunoglobulins, but the exact relation remained obscure. The 1959 finding that immunoglobulins contained multiple polypeptide chains suggested that Bence Jones proteins were homogeneous light chains of the myeloma protein made by the tumor but not incorporated into whole molecules. This hypothesis was confirmed by my student Joseph A. Gally and me in 1962. Because

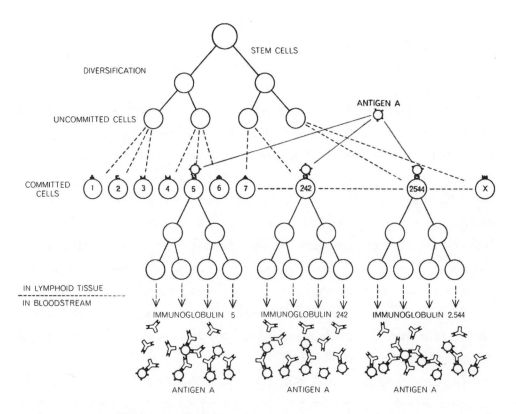

SELECTIVE-IMMUNITY THEORY holds that stem cells (precursors of antibody-producing cells) contain information for making all possible antibodies; at some point in embryonic development each is committed to producing a unique immunoglobulin (*numbers*). These receptor antibodies can interact with various antigens. A single antigen may be recognized by more than one antibody-producing cell. Interaction of an antigen with a "ell stimulates proliferation of the cell and the synthesis of antibody.

different Bence Jones proteins had different amino acid compositions, we compared their properties with those of light chains of antibodies. These comparisons were instrumental in suggesting that antibodies with different antigen specificities differ from one another in the sequence of amino acids of which they are composed. Moreover, because the light chains produced by the myeloma tumors were pure, available in large amounts and smaller than a whole immunoglobulin molecule, it became possible to study the details of their chemical structure.

Once it had become apparent that these proteins might provide a clue to the nature of antibody variability, a number of laboratories undertook the task of determining their exact amino acid sequence. Since no two individuals produce the same Bence Jones proteins, each laboratory reported a different sequence, but from the first report of partial sequences by Norbert G. D. Hilschmann and Lyman C. Craig at Rockefeller University in 1965 it became clear that these proteins had a singular structure. Each molecule contained about 214 amino acids, linked together in a polypeptide chain. (The amino acids are numbered starting from the end of the molecule that is made first, the amino terminus.) From position 109 on various Bence Jones proteins had essentially the same sequence, and accordingly this part of the molecule was called the constant region. In striking contrast, the sequence of the first 108 amino acids differed markedly from one Bence Jones protein to another, and this first part of the chain was designated the variable region. Concurrent studies in several laboratories suggested that the heavy chains of myeloma proteins also had variable and constant regions. The homogeneity of the constant region enabled Robert L. Hill and his associates at Duke University to determine the amino acid sequence of the Fc fragment of rabbit immunoglobulin.

It was against this background that my colleagues and I decided to attempt the determination of the complete amino acid sequence of a whole immunoglobulin molecule. The earlier structural studies had suggested an overall picture and we wanted to confirm and extend it in detail so that we could apply it to an analysis of the origin and function of antibodies. We obtained a large amount of plasma from a patient with multiple myeloma, because it was essential to have enough of at least one myeloma protein. As a matter of fact, we obtained two different proteins from different patients for purposes of comparison. The difficulties of our project were related to the enormous size of the molecule, which has 19,996 atoms and is larger in terms of the number of unique amino acid sequences than any protein that had been determined up to that time. Our approach was based on the pioneering methods first developed by Frederick Sanger to analyze the insulin molecule, and the challenge, successfully met last year, was whether these methods would suffice.

The immunoglobulin molecule is about 25 times as large as insulin. If

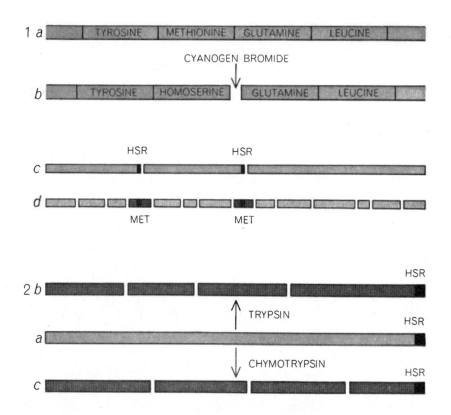

ANTIBODY CHAIN is cleaved by cyanogen bromide (CNBr) and the resulting fragments are ordered (1). The CNBr breaks a chain (a) at methionyl residues, which are converted into homoserine (b). To order CNBr fragments (c) the original chain is cleaved also with the enzyme trypsin, the tryptic fragments containing methionine are isolated (d) and their sequences are compared with those of the ends of the CNBr fragments. Then the amino acid sequence of each CNBr fragment must be determined (2). This is done by cleaving a CNBr fragment (a) with trypsin and determining the sequence of each tryptic peptide (b) by a chemical procedure. The tryptic peptides are ordered by comparison with the composition and partial sequences of different peptides (c) made by cleaving with chymotrypsin.

we could break the immunoglobulin molecule, its chains and its fragments into small pieces about the size of the insulin molecule itself, then we could use standard methods for determining amino acid sequence. A useful tool for such protein surgery had already been devised by Erhard Gross and Bernhard Witkop of the National Institutes of Health. It depended on cyanogen bromide, a reagent that selectively cleaves polypeptide chains at the positions occupied by the sulfur-containing amino acid methionine; because there were just a small number of methionines in the molecule, we could expect a decently small number of pieces. With this reagent we were able to cleave the heavy chain of the immunoglobulin into seven pieces and the light chain into three pieces. Each piece was then separated from the others by chromatography. A key procedure in these separations was molecular sieving on Sephadex, a technique developed largely by Jerker O. Porath of the University of Uppsala, which speeded up the thousands of separations required to determine the structure of immunoglobulin.

After the fractionation of the pieces the next step was to establish their order. This was done by cleaving the original chains not with cyanogen

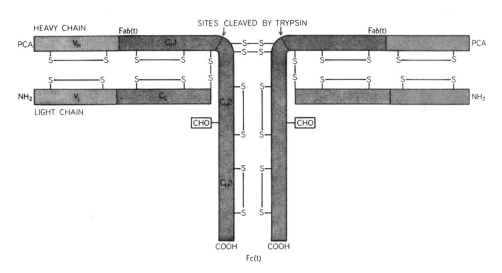

STRUCTURE of the immunoglobulin determined by the author and his colleagues shows two kinds of chain and regions in each. The protein can be cleaved into two antibody binding fragments, Fab(t), and a "crystallizable" fragment, Fc(t). Sulfur-sulfur bonds are designated—S—S—. Light chains have variable and constant regions ($V_L$ and $C_L$). Heavy chains have a variable region ($V_H$) and a constant region divisible into three homology regions ($C_H1$, $C_H2$ and $C_H3$). CHO indicates carbohydrate. Chains have amino ($NH_2$), carboxyl ($COOH$) and pyrollidonecarboxylic acid ($PCA$) ends. $V_{II}$ and $V_I$ are homologous, as are $C_L$, $C_H1$, $C_H2$ and $C_H13$.

AMINO ACID SEQUENCES of the variable regions (*top*) and of the constant homology regions (*bottom*) were fully determined. The extent of the homology between the two variable regions and among the four constant homology regions is indicated by the coloring or shading of identical residues in each position; dark and light shading indicates identities that occur in pairs at one position. Gaps have been introduced to maximize the homology. The numbering across the top is that of positions of residues in light chains.

bromide but with enzymes that attack polypeptides at other specific sites. Those peptides that contained methionine and that would therefore overlap the cyanogen bromide fragments were then isolated. By comparing the two kinds of fragment we could see which ends of the cyanogen bromide fragments butted up against one another.

Each separate cyanogen bromide fragment could now be studied independently as if it were a separate small protein or polypeptide. Accordingly it was cleaved with enzymes into smaller peptides, which were separated. When small pure peptides were obtained, the sequence of their amino acids was determined directly by a chemical procedure. The order of the peptides was then established by breaking the whole cyanogen bromide fragment with a second enzyme that cleaved it at different sites and isolating a second set of peptides that overlapped the first set.

The sequence determination was thus a "two pass" procedure. In the first pass we obtained cyanogen bromide fragments and ordered them. In the second we treated each fragment as a separate protein, obtaining its peptides, ordering them and determining their amino acid sequence. When these tasks were finished, there remained the job of determining the location of the bonds between the sulfur atoms of the amino acid cysteine that helped to link the chains and parts of the chains together.

The completed structure showed that the antibody molecule differed from proteins that had been analyzed earlier not only in size but also in more unusual ways. Our molecule was what is classified as a γG, or gamma G, immunoglobulin molecule, an example of the most prevalent class of immunoglobulins. As earlier studies had suggested, such a molecule consists of two identical light and two identical heavy chains. The structure is symmetrical, each half consisting of one light and one heavy chain. Although the actual shape or three-dimensional structure of the chains is not known, it is established that they are held together by weak forces and by interchain sulfur-sulfur bonds between corresponding pairs of cysteines; similar intrachain bonds are formed within each chain at approximately equal intervals. The most striking feature of the structure is its division into two kinds of region, variable regions and constant regions, whose disposition is related to these intervals. The length of the variable regions was determined by comparing the amino acid sequences of the light and the heavy chains to the sequences of Bence Jones proteins and to the sequence of another heavy chain that was analyzed concurrently. As in the Bence Jones proteins, the regions are so named because in different antibodies the variable regions differ in the sequence of amino acids that make up the chain, whereas the constant regions have the same sequence in each of the major classes of antibodies (except for a single variable amino acid at position 191). It has now been firmly established that it is the different sequences in the variable regions that give different shapes to various antigen-binding sites. The variety of shapes provides for a range of specific interactions with a great variety of antigens, including small molecules, other proteins, carbohydrates and even DNA itself.

There is another feature of the structure that merits notice. Detailed examination of the constant regions showed evidence of internal periodicity,

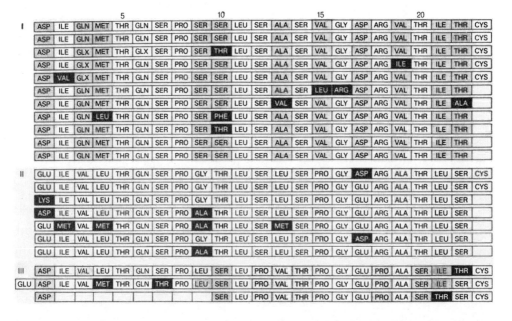

PATTERN OF VARIATION of three different subgroups of one class of light-chain variable regions yields clues to their genetic origin. Each line represents a partial sequence (the first 23 residues) determined in the laboratories of H. D. Niall and P. Edman, C. Milstein, Norbert G. D. Hilschmann, Frank W. Putnam or Lee Hood. Each subgroup (*roman numerals*) has a characteristic sequence, indicated in each case by a dominant color or shade of gray. Within each subgroup there are variations (*black*) that arose from mutations. (GLX refers to positions where it was not yet definitely established whether the residue was glutamic acid or glutamine.)

which had already been hinted at by the distribution of the sulfur-sulfur bonds. Portions of the constant region of the heavy chain turn out to have homologous amino acid sequences, that is, sequences more similar than could occur by chance. These portions are designated $C_H1$, $C_H2$ and $C_H3$, and each is homologous also to the constant region of the light chain, $C_L$. It is these constant regions that carry out functions of the molecule other than the binding of antigens. For example, $C_H2$ is believed to be bound by members of a complex family of serum proteins known as complement, thus beginning the series of reactions that is capable of killing cells, one of the aspects of the immune response.

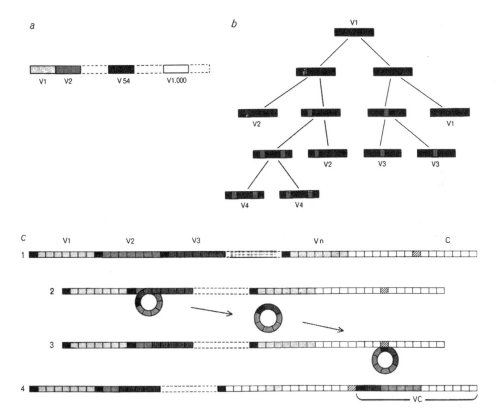

DIVERSITY of variable regions can be explained by three theories. A large number of genes, one for each variable region, could have arisen in the course of evolution (a). Alternatively, there could be one V gene, which mutates in an individual animal's body cells during development to produce the required variety (b). Finally, several V genes could evolve by mutation and be selected during evolution, and then be recombined in many different ways in the animal (c). In this process the evolved genes (1) might recombine to form a ring-shaped V-gene episome (2), a variant gene composed of sequences from adjacent V genes. The episome might be translocated and become integrated with the C gene (3) to form the complete VC gene that is expressed (4).

The homology of the constant regions and the somewhat weaker homology of the variable regions to one another is demonstrated by directly comparing their amino acid sequences. This is an unusual finding, and it means that the regions must be related in their evolutionary origins. It is likely that present-day antibodies have evolved by a process known as gene duplication. A primitive gene of a size sufficient to specify one homology region must have doubled and tripled, thereby forming a larger gene whose segments then became somewhat different from one another, as reflected in the sequences. By a similar process the genes for the two kinds of chain, heavy and light, appear to have had a common ancestor. This hypothesis, which was first suggested by Hill at Duke and S. Jonathan Singer of the University of California at San Diego, is strongly supported by the structure of the whole molecule.

A comparison of the variable regions with the four constant homology regions shows that although they have roughly the same length, they have few sequences in common. Did they also arise from the same original gene? Probably, far back in evolution, but if so, they must have diverged rapidly as they carried out different functions of the antibody. Indeed, studies by C. Milstein of the Medical Research Council in England and by Lee Hood of the National Institutes of Health indicate that in each individual there must be more than one gene for each variable (V) region. Earlier, pioneering genetic investigations by Jacques Oudin of the Pasteur Institute and by Rune Grubb of the University of Lund had laid the groundwork for the conclusion that there is only one gene for each constant (C) region. Since the polypeptide chains of antibodies appear to be made in one piece, as are other proteins, it seems that information from two genes is required to specify a single polypeptide chain. This is a unique situation, because in all proteins that have so far been investigated a single gene is enough to specify a single polypeptide chain.

The analysis of antibodies, then, poses two special problems: How can the V genes vary so that many different V regions are made in each individual? And how can such a V gene, which evolved to give the antibody system a range of different combining sites, be joined with a C gene that evolved to specify the constant portions of the chains and thus carry out effector functions?

Before attempting to suggest answers to these questions, let us look at the actual variation seen in V regions. The variation has several important characteristics. First, the genetic-code dictionary (in which each amino acid is coded for by a triplet of three DNA nucleotides) reveals that the variation arises from one-base changes in the code words for the amino acids in each variable position. This means that the variations were caused by mutations, just as in the case of other proteins. Second, not every position in the V region varies. For example, no one has ever observed that any of the cysteines that contribute to the sulfur bonds are missing or replaced by another amino acid. Third, certain positions seem to have more variations than others, although the number of examples is still too small for one to be completely sure of this. These last two observations mean that the variation is not random but is the result of some kind of selection. We can conclude that, as in other proteins, both mutation and selection are responsible for antibody variation.

The question about the origin of variability can be resolved into two more pointed questions. They are: Where and when do these processes occur? How many V genes are required? One theory suggests that the variation and selection occur during evolution, so that in an animal each different V region has a corresponding V gene. This would require a very large number of V genes in the germ cells. Another theory states that there is one V gene, which mutates not during evolution but somatically—in the body cells of the individual animal—and that the mutant cells are somehow selected. A third theory, which I favor, is that there are a few V genes, which have mutated and have been selected in evolution but which then recombine somatically in the cells of the animal to provide the broad variational pattern. This last theory has the advantage that the same processes that recombine the V genes could also accomplish the fusion of V and C genes that is required to make a single antibody chain; one can thus account with one mechanism for the two questions: How is antibody diversity created? How are V and C genes joined?

The mechanism may be one that is somewhat similar to mechanisms that have already been described for infection of the bacterium *Escherichia coli* by the bacterial virus lambda. A piece of DNA (a V gene) could be removed from a row of V genes (each having evolved to be slightly different) and could then be inserted and fused with a C gene. If the DNA is removed as a ring, the process would effectively permute the sequence, leading to variation. As mentioned above, the alternative somatic theory is that a single V gene is mutated and then translocated, following which the cell making the VC

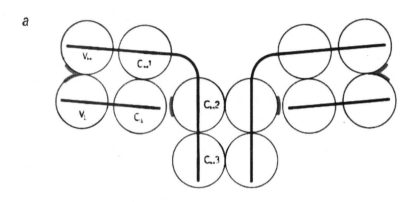

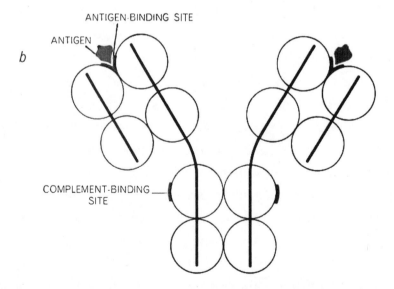

BINDING OF ANTIGEN may stabilize a change in the shape of the antibody that triggers a series of immune reactions. Here the antibody is drawn as a flexible grouping of compact domains, with the chain structure suggested by the heavy black lines (*a*). The antigen is bound by the variable regions, perhaps facilitating a pivoting movement of the molecule that exposes effector sites, such as the complement-binding site, in the C regions (*b*).

product is selected for or against within the body. Which of these theories is correct remains to be defined, but both theories require a process of assembly of a VC gene from separate V and C genes. This requirement suggests that it may be by the translocation of genes that the cell achieves its goal of making just one kind of immunoglobulin. Molecular differentiation of this type is so far unique to the immune system, but it may in fact turn out to be important in other systems of differentiation among higher organisms.

Although the molecular details of the mechanism of translocation and recombination remain hypothetical, there is some recent evidence that we are on the right track in concluding that antibody variation is somatic in origin. The evidence comes from studies of the genetics of mouse immunoglobulins done with my student Paul D. Gottlieb. These experiments are still in their early stages, however, and the actual mechanism of variation has so far not been demonstrated.

The origin of the required diversity and the restriction of that diversity to one kind of antibody for each cell seem to be mirrored in the structure of the antibody molecule. What does antibody structure tell us about how these molecules actually carry out their functions? It is clear that antibodies have two kinds of task: first, recognizing the antigen, and second, doing something to the antigen or initiating a chain of cellular responses. The task of antigen-binding is delegated to the V regions. The more dynamic role of influencing cellular responses, the binding of complement and the initiation of processes that alter the antigen appears to be the function of the C regions. In some sense the antibody molecule must behave as a switch: binding the antigen must change the antibody's state in such a way as to "turn on" its effector functions. It is known that protein molecules can act as switches by changing their shape. There is now a hint that this may be the case for antibodies. When antibodies are viewed in the electron microscope after combination with antigens, their dimensions appear to be smaller than those of unbound immunoglobulins measured by X-ray scattering in solution. This raises the possibility that the binding of the antigen causes a rearrangement of the structure of the antibody molecule, which is known to be somewhat flexible. The rearrangement might consist, for example, of a pivoting movement involving part of the constant region of the heavy chains. Binding sites for complement might be exposed by this pivoting, as well as other sites for different effector functions. Similar mechanisms may be involved in triggering the antibody-producing cell to divide and mature.

A s these hypotheses indicate, much remains to be done in the field of antibody structure. What has been learned indicates that the antibody system is special and that it may have evolved to solve its problem of molecular recognition in a unique way. There remains the intriguing possibility, however, that the special genetic mechanisms hinted at by the differentiation of antibody-producing cells will be found in other systems of cellular differentiation; certainly there are at least conceptual similarities to some other systems of pattern recognition, such as those of the central nervous system. In any event, whether the immune system turns out to be unique or representative of a more general type of evolutionary development, we can expect practical consequences of great significance for fields of study such as immune tolerance, organ transplantation and autoimmune disease to flow from a continuing analysis of the structure of antibodies.

## ABOUT THE AUTHOR

**Gerald M. Edelman** (born 1929) was awarded the Nobel Prize in Physiology or Medicine in 1972 along with R. R. Porter for their discoveries concerning the chemical structure of antibodies. Dr. Edelman is Director of The Neurosciences Institute and President of Neurosciences Research Foundation, the publicly supported not-for-profit organization that is the Institute's parent.

# The Immune System

*This diffuse organ has the assignment*
*of monitoring the identity of the body.*
*Its basic constituents are lymphocytes and*
*antibody molecules, which recognize both*
*foreign molecules and one another.*

## Niels Kaj Jerne

The immune system is comparable in the complexity of its functions to the nervous system. Both systems are diffuse organs that are dispersed through most of the tissues of the body. In man the immune system weighs about two pounds. It consists of about a trillion ($10^{12}$) cells called lymphocytes and about 100 million trillion ($10^{20}$) molecules called antibodies that are produced and secreted by the lymphocytes. The special capability of the immune system is pattern recognition and its assignment is to patrol the body and guard its identity.

The cells and molecules of the immune system reach most tissues through the bloodstream, entering the tissues by penetrating the walls of the capillaries. After moving about they make their way to a return vascular system of their own, the lymphatic system. The tree of lymphatic vessels collects lymphocytes and antibodies, along with other cells and molecules and the interstitial fluid that bathes all the body's tissues, and pours its contents back into the bloodstream by joining the subclavian veins behind the collarbone. Lymphocytes are found in high concentrations in the lymph nodes, way stations along the lymphatic vessels, and at the sites where they are manufactured and processed: the bone marrow, the thymus and the spleen.

The immune system is subject to continuous decay and renewal. During the few moments it took you to read this far your body produced 10 million new lymphocytes and a million billion new antibody molecules. This might not be so astonishing if all these antibody molecules were identical. They are not. Millions of different molecules are required to cope with the task of pattern recognition, just as millions of different keys are required to fit millions of different locks.

The specific patterns that are recognized by antibody molecules are epitopes: patches on the surface of large molecules such as proteins, polysaccharides and nucleic acids. Molecules that display epitopes are called antigens. It is hardly possible to name a large molecule that is not an antigen. Let us consider protein molecules, which include enzymes, hormones, transport molecules such as hemoglobin and the great variety of molecules that are incorporated in cellular membranes or form the outer coat of viruses or bacteria.

## Antigens and Antibodies

E ach of the innumerable protein molecules is made up of polypeptide chains: linear strings of a few hundred amino acids chosen from a set of 20 amino acids. The number of amino acids in a large protein molecule is about equal to the number of letters in the column of text you are now reading, which is a linear string of letters chosen from an alphabet of 26 letters. Different protein molecules have different amino acid sequences just as different texts have different letter sequences. The string of letters in this column of text has been neatly "folded" into successive lines. The polypeptide chains of a protein molecule are also folded, although not so neatly. Their structure looks more like what you would obtain by haphazardly compressing a few yards of rope between your hands. There is nothing haphazard, however, about the folding of a particular polypeptide chain; the folding, and thus the ultimate conformation of the protein molecule, is precisely dictated by the amino acid sequence.

The parts of the folded chains that lie at the surface of a protein molecule make up its surface relief. An epitope (or "antigenic determinant") is a very small patch of this surface: about 10 amino acids may contribute to the pattern of the epitope. As Emanuel Margoliash of the Abbott Laboratories and Alfred Nisonoff of the University of Illinois College of Medicine showed for different molecules of cytochrome *c*, the replacement of just one amino acid by another in a polypeptide chain of a protein frequently leads to the display of a different epitope. The immune system recognizes that difference and is able to check on mutant cells that make mistakes in protein synthesis. Not only can an individual immune system recognize epitopes on any protein or other antigen produced by any of the millions of species of animals, plants and microorganisms but also it can distinguish "foreign" epitopes from epitopes that belong to the molecules of its own body. This recognition is a crucial event, since antibody molecules attach to the epitopes they recognize and thereby earmark the antigens (or the cells that carry them) for destruction or removal by other mechanisms available to the body.

Epitopes are recognized by the combining sites of antibody molecules. An antibody is itself a protein molecule consisting of more than 20,000 atoms. It is made up of four polypeptide chains: two identical light chains and two identical heavy chains. A light chain consists of 214 amino acids and a heavy chain of about twice as many. Antibody molecules are alike except for the amino acids at about 50 "variable" positions among the first 110 positions, which constitute what is called the variable region of both the light and the heavy chains. At the tip of each variable region there is a concave combining site whose three-dimensional relief enables it to recognize a complementary epitope and make the antibody molecule stick to the molecule displaying that epitope. Whether a combining site will recognize one epitope or a different one depends on which amino acids are located at the variable positions. If at each of 50 positions of both chains there were an independent choice between just two amino acids, there would be $2^{100}$ (or $10^{30}$) potentially different molecules! The situation is not that simple, however. The chains fall into subgroups, within each of which there are far fewer than 50 variable positions. On the other hand, at some of those variable positions, clustered in so-called hot spots, the choice is actually among more than two alternative amino acids. There is general agreement that the differences in amino acid sequence among antibody molecules derive from mutations that have occurred in the genes encoding antibody structure.

## The Recognition Problem

Smallpox being the nasty disease it is, one might expect nature to have designed antibody molecules with combining sites that specifically recognize the epitopes on smallpox virus. Nature differs from technology in its approach to problem solving, however: it thinks nothing of wastefulness. (For example, rather than improving the chance that a spermatozoon will meet an egg cell, nature finds it easier to produce millions of spermatozoa.) Instead of designing antibody molecules to fit the smallpox virus and other noxious agents, it is easier to make millions of different antibody molecules, some of which may fit. By way of analogy, suppose someone makes gloves in 1,000 different sizes and shapes: he would have a sufficiently well-fitting glove for almost any hand. Now imagine that hands were a great deal more variable; for example, the length of the fingers on a hand might vary independently from one inch to six inches. By making, say, 10 million gloves of different shapes the manufacturer would nevertheless be able to fit practically any hand—at the expense of efficiency, to be sure, since most of the gloves might never find a customer to fit them. Now be more wasteful still: have a factory with machines capable of turning out gloves of a billion different shapes, but turn off 99 percent of the machines, so that the factory actually turns out a random collection of 10 million of the potential billion shapes. You would still be doing all right. So would your colleague running a similar factory. Although the two sets of gloves you and he would make would show only a 1 percent overlap, each set would serve its purpose well enough.

That is how some of us think the immune system solves its recognition problem. By a more or less random replacement of amino acids in the hot-spot positions of the variable regions of antibody polypeptide chains, a set of millions of antibody molecules is generated with different combining sites that will fit practically any epitope well enough. As has been demonstrated by Jacques Oudin of the Pasteur Institute and by Andrew Kelus and Philipp G. H. Gell of the University of Birmingham for rabbits and by Brigitte A. Askonas, Allan Williamson, Brian Wright and Wolfgang Kreth of the National Institute for Medical Research in London for mice, individual animals make use of entirely different sets of antibodies capable of recognizing a given epitope.

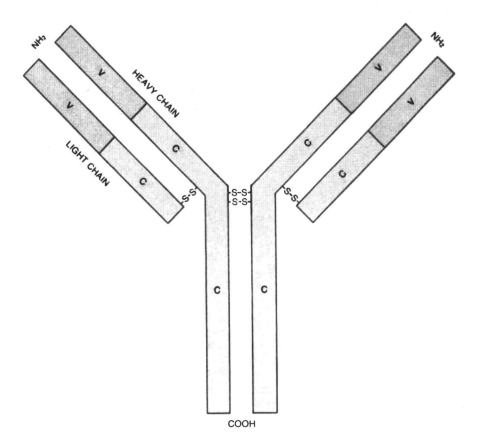

LINEAR STRUCTURE of an antibody molecule is shown schematically. The two heavy chains and two light ones are connected by disulfide bridges. Each chain has an amino end (NH₂) and a carboxyl end (COOH). Chains are divided into variable (V) regions, in which the amino acid sequence varies in different antibodies, and constant (C) regions.

There is one serious snag in all of this, to which I alluded above: one's immune system does not seem to recognize the epitopes on molecules and cells that are part of one's own body. This property, which Sir Macfarlane Burnet called the discrimination between self and not-self, is often referred to as self-tolerance. You might think that self-tolerance derived from nature's being wise enough to construct the genes coding for your antibodies in such a way as not to give rise to combining sites that would fit epitopes occurring in your own body. It can easily be shown, however, that this is not so. For example, your father's antibodies could recognize epitopes occurring in your mother; some antibody genes inherited from your father should therefore code for antibodies recognizing epitopes inherited from your mother.

Self-tolerance, then, is not innate. It is something the immune system "learned" in embryonic life by either eliminating or "paralyzing" all lymphocytes that would produce self-recognizing antibodies. An original observation of this phenomenon by Ray D. Owen of the California Institute of Technology was generalized in a theoretical framework by Burnet and received experimental confirmation by P. B. Medawar in the 1950's, bringing Nobel prizes to Burnet and Medawar in 1960.

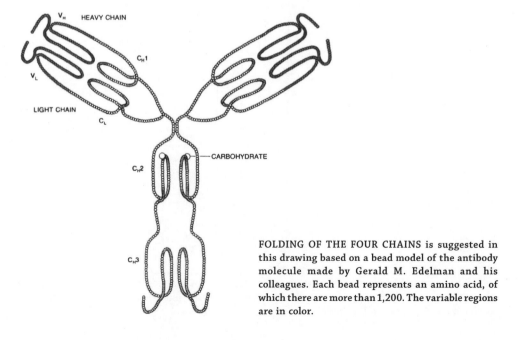

FOLDING OF THE FOUR CHAINS is suggested in this drawing based on a bead model of the antibody molecule made by Gerald M. Edelman and his colleagues. Each bead represents an amino acid, of which there are more than 1,200. The variable regions are in color.

## The Lymphocyte

Emil von Behring and Shibasaburo Kitazato discovered the existence of antibodies in Germany in 1890, but it was not until the 1960's that the structure of antibodies was determined, through the investigations initiated by R. R. Porter of the University of Oxford and Gerald M. Edelman of Rockefeller University [see "The Structure of Antibodies," by R. R. Porter, SCIENTIFIC AMERICAN, October, 1967, and "The Structure and Function of Antibodies," by Gerald M. Edelman, SCIENTIFIC AMERICAN, August, 1970]. The two men shared a Nobel prize last year for that work. Long before the structure of antibodies was known, however, antibodies had been the subject of detailed studies. And yet it was not known that antibodies are produced by activated lymphocytes. Even 20 years ago lymphocytes were not thought to have anything to do with the immune system, something that seems odd now that they are known to constitute the immune system! It was only in the early 1960's that the involvement of lymphocytes was proved by James L. Gowans and Douglas McGregor of the University of Oxford.

Most lymphocytes (about 98 percent of them) do not actually secrete antibody. They are the "small" lymphocytes, spherical cells measuring about a hundredth of a millimeter in diameter, and they are said to be in a resting state. In order to secrete antibody a small lymphocyte must first become enlarged. In that state it can not only secrete antibody molecules but also divide and become two cells, which in turn can become four cells and so on. The offspring cells constitute the clone, or cell line, derived from one small lymphocyte.

As was originally postulated by Burnet in 1957, the antibody molecules produced by a lymphocyte and by the cells of its clone all have identical combining sites [see "The Mechanism of Immunity," by Sir Macfarlane Burnet; SCIENTIFIC AMERICAN, January, 1961]. G. J. V. Nossal, Burnet's successor as director of the Walter and Eliza Hall Institute of Medical Research in Melbourne, and his co-workers have accumulated much of the experimental evidence that now firmly supports this "single commitment" of the lymphocyte [see "How Cells Make Antibodies," by G. J. V. Nossal; SCIENTIFIC AMERICAN, December, 1964]. The cells of one lymphocyte clone are committed to the expression of two particular genes coding for particular variants of the variable regions of the light chain and the heavy chain. Already in its resting, non-secreting state a small lymphocyte produces a relatively small number of its particular antibody molecules, which it displays on the surface of its outer membrane. These antibody molecules are the "receptors" of the cell. A small lymphocyte displays about 100,000 receptors with identical combining sites, which are waiting, so to speak, for an encounter with an epitope that fits them.

When such an epitope makes contact, the lymphocyte can either become "stimulated" (respond positively) or become "paralyzed" (respond negatively), which is to say it is no longer capable of being stimulated. Investigations in progress by David S. Rowe of the World Health Organization, working in Lausanne, and Benvenuto Pernis at our Basel Institute for Immunology suggest that the distinction between excitatory and inhibitory signals may reside in differences in the constant regions of the lymphocyte's receptor antibody molecules. Whether a lymphocyte will choose to respond positively or negatively can be shown to depend on several conditions: the concen-

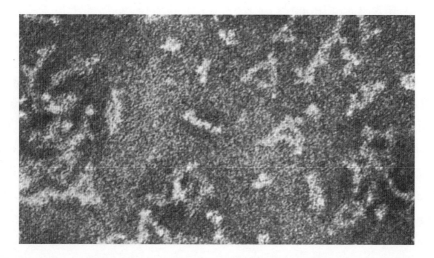

ANTIBODY MOLECULES are visible in an electron microscope when they are linked to antigens and one another in antigen-antibody complexes. In this micrograph, made by N. M. Green and the late Robin Valentine of the National Institute for Medical Research in London, rabbit antibodies are enlarged 500,000 diameters. The antigen is a short polypeptide chain with a dinitrophenyl group at each end; the antibodies are from a rabbit that was immunized against dinitrophenyl epitopes. The antigens (too small to be visible) link antibodies to form polygonal complexes whose geometry derives from antibody structure.

tration of the recognized epitopes, the degree to which those epitopes fit the combining sites of the receptors, the way the epitopes are presented (for example whether they are presented on molecules or on cell surfaces) and the presence or absence of other lymphocytes that can "help" or "suppress" a response. Much current experimentation aims at clarifying these complex matters.

A stimulated lymphocyte faces two tasks: it must produce antibody molecules for secretion and it must divide in order to expand into a clone of progeny cells representing its commitment. Progeny cells that go all out into the production and secretion of antibody molecules are called plasma cells. Each of them must transcribe its antibody genes into 20,000 messenger-RNA molecules that serve 200,000 ribosomes, enabling the cell to produce and secrete 2,000 identical antibody molecules per second. Other cells of the clone do not go that far; they revert to the resting state and represent the "memory" of the occurrence, ready to respond if the epitope should reappear. The immunological memory of what Stephen Fazekas de St. Groth of the University of Sydney, who is now working in our laboratory in Basel, has called "original antigenic sin" is remarkably persistent. People who are now 90 years old, for example, and had influenza in the 1890's still possess circulating antibodies to the epitopes of the influenza virus strains that were prevalent at that time.

If a lymphocyte that recognizes an epitope does not become stimulated, it may become paralyzed. Paralysis can occur when a lymphocyte is confronted by very high concentrations of epitope; this is called high-zone tolerance. David W. Dresser and N. Avrion Mitchison, who were working at the National Institute for Medical Research, have shown that paralysis can also result from the continuous presence of extremely small epitope concentrations, below the threshold required for stimulation; this is called low-zone tolerance. We need more knowledge of the mechanisms leading to paralysis, not only in order to understand how the immune system learns to tolerate self-epitopes but also to be able to induce the system to tolerate organ transplants.

## Germ-Line and Soma Theories

The enormous diversity of antibodies raises the question of the origin of the genes that code for the variable regions of antibody molecules. Essentially two answers have been proposed to this question. They are the germ-line theory and the somatic theory. The argument of the germ-line theory is straightforward: All the cells of the body, including lymphocytes, have the same set of genes, namely those in the fertilized egg from which the individual arose. Therefore genes for any antibody that an individual can make must already have been present in the fertilized egg cell. They are all transmitted to the individual's children through the germ-cell line: egg and spermatozoa and their precursors.

The somatic theory does not accept this approach. It is argued that the immune system needs millions of different antibodies for epitope recognition. Individual mice of an inbred strain, all having the same germ-line genes, have been shown to make use of entirely different sets of antibody molecules. The germ-line theory implies that the set of all these sets is

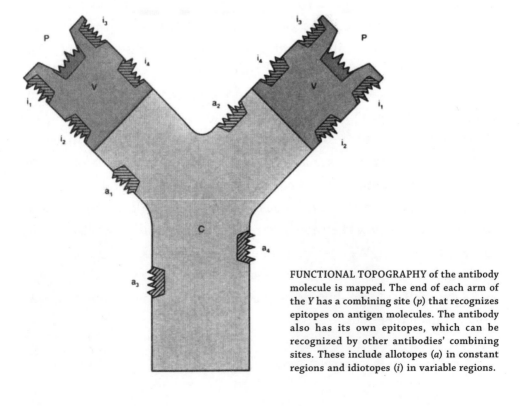

FUNCTIONAL TOPOGRAPHY of the antibody molecule is mapped. The end of each arm of the Y has a combining site (*p*) that recognizes epitopes on antigen molecules. The antibody also has its own epitopes, which can be recognized by other antibodies' combining sites. These include allotopes (*a*) in constant regions and idiotopes (*i*) in variable regions.

represented in the genes of every single mouse of that strain. In that case, however, many of the genes would seem to have no survival value for the mouse, so that such a large number of genes cannot arise or be maintained in Darwinian evolution. Most antibody genes must therefore have arisen in the course of the somatic development of the individual by modification of a smaller number of germ-line genes. That is the point of departure for several variants of the somatic theory.

I have proposed that an inherited set of germ-line genes code for antibodies against certain self-epitopes. The clones of cells expressing these genes become suppressed except for mutant cells that, by an amino acid replacement, display new combining sites on their antibody receptor molecules. These mutant cells represent the enormous repertoire of antibodies that recognize foreign epitopes. An organ that could breed such mutant cells is the thymus gland. More than $10^{10}$ new lymphocytes arise in the

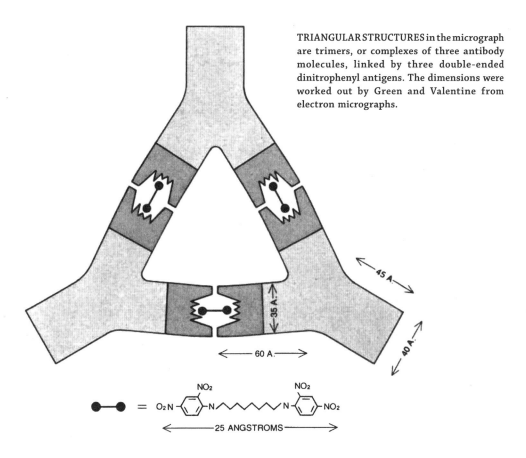

TRIANGULAR STRUCTURES in the micrograph are trimers, or complexes of three antibody molecules, linked by three double-ended dinitrophenyl antigens. The dimensions were worked out by Green and Valentine from electron micrographs.

thymus every day; the vast majority of these cells are killed in the thymus or immediately after they leave it.

It is not possible here to discuss the merits of these theories. That would require consideration of a large body of experimental results, such as the explorations of the genetics of immune responsiveness by Baruj Benacerraf of the Harvard Medical School, Hugh O. Mc-Devitt of the Stanford University Medical Center and Michael Sela of the Weiz-mann Institute of Science in Israel.

## *T* Cell and *B* Cell

All the lymphocytes that circulate in the tissues have arisen from precursor cells in the bone marrow. About half of these lymphocytes, the *T* cells, have passed through the thymus on their way to the tissues; the other half, the *B* cells, have not. This dichotomy was first discovered by Henry N. Claman of the University of Colorado Medical School and was characterized by Jacques F. A. P. Miller and Graham Mitchell, both of whom are now working with US in Basel. It has been the subject of thousands of investigations during the past five years. *T* cells and *B* cells cannot be distinguished by their form. Only *B* cells and their progeny cells secrete antibody molecules. One might think that this leaves little scope for T-cell function. On the contrary, *T* cells appear to be all-important. They too can recognize epitopes and must therefore, almost by definition, possess antibody molecules as surface receptors, although these receptor molecules have been much harder to demonstrate experimentally than those on *B* cells.

*T* cells can kill other cells, such as cancer cells, and transplanted tissues that display foreign epitopes. *T* cells can also suppress *B* cells or alternatively can help *B* cells to become stimulated by epitopes. This "helper" function of *T* cells has been repeatedly demonstrated both in animal experiments and in experiments with cells in culture. In the cell-culture experiments, based on a technique developed by Richard W. Dutton and Robert 1. Mishell at the University of California at San Diego, lymphocytes taken from the spleen of an untreated animal are grown in a plastic dish together with molecules or cells that display foreign epitopes. After a few days' incubation lymphocytes that produce and secrete antibody molecules against the foreign epitopes can be shown to be present in the culture by the assay method for single antibody-producing cells. These antibody molecules are made by *B* cells, but the experiment will not work if only *B* cells are present. As soon as *T* cells are added to the culture dish, however, the *B* cells begin to respond and to produce antibody.

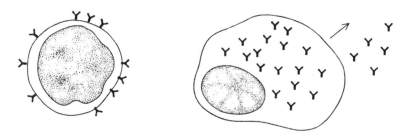

LYMPHOCYTES, the cells of the immune system, produce antibodies. Each cell is committed in advance to the production of one specific antibody. In its resting state, as a small lymphocyte (*left*), the cell displays such antibody molecules on its surface as "receptors." The advent of an antigen with an epitope that fits the combining site of this particular antibody molecule may stimulate the lymphocyte to grow, change in structure and divide, eventually giving rise to a large number of plasma cells (*right*): lymphocytes specialized for the rapid synthesis and secretion of this cell line's specific antibody molecules.

The dichotomy of the immune system into *T* and *B* lymphocytes adds a further dimension to the conceptual framework needed for the system's comprehension. That is not only an intellectual need but also a practical one, since the immune system is now known to be crucially involved in a vast number of diseases ranging from microbial infections and allergies to cancer, rheumatism, auto-immunity and many other degenerative disorders of aging.

## The Lymphocyte Network

I have mentioned two striking dualisms within the immune system. One is the dichotomy of the lymphocytes into *T* cells and B cells, with functions that are partly synergistic and partly antagonistic. The second is the duality of the potential response of a lymphocyte when its receptors recognize an epitope: it can either respond positively (become stimulated) or respond negatively (become paralyzed). It is important to realize that the immune system displays a third dualism, namely that antibody molecules can recognize and can also be recognized. They not only have combining sites enabling them to recognize epitopes but also display epitopes enabling them to be recognized by the combining sites of other antibody molecules. That is true for the antibody molecules attached to the outer membranes of lymphocytes and serving as receptors as well as for the freely circulating antibody molecules, which can be regarded as messages released by lymphocytes.

Epitopes occur on both the constant and the variable regions of an antibody molecule. Since the patterns of the variable-region epitopes are determined by the variable amino acid sequences of the polypeptide chains, there are millions of different epitopes. The set of such epitopes on a given antibody molecule was named the idiotype of that molecule by Oudin. When antibodies produced by animal A are injected into animal B, animal B will produce antibodies against the idiotypic epitopes ("idiotopes") of the injected antibody molecules. That is also true when A and B belong to the same animal species and even when they are of the same inbred strain, that is, when they are genetically identical. Evidence is emerging that, within one animal, the idiotopes occurring on one antibody molecule are recognized by combining sites on a set of other antibody molecules, and that the idiotopes on the receptor molecules of one lymphocyte are recognized by the combining sites of the receptor molecules of a set of other lymphocytes. We thus have a network of lymphocytes and antibody molecules that recognize other lymphocytes and antibody molecules, which in turn recognize still others.

I am convinced that the description of the immune system as a functional network of lymphocytes and antibody molecules is essential to its understanding, and that the network as a whole functions in a way that is peculiar to and characteristic of the internal interactions of the elements of the immune system itself: it displays what I call an eigen-behavior. (Eigen in German means peculiar to, or characteristic of. Eigen-behavior is analogous to such concepts as the eigenvalue or eigenfrequency of certain physical systems.) There is an increasing body of evidence for this view.

Antibody molecules are normally present in the blood in a concentration of about $5 \times 10^{16}$ molecules per milliliter. The total concentration of combining sites and idiotopes is therefore of the order of $10^{17}$ per milliliter. If the immune system made use of 10 million different combining sites and 10 million different idiotopes, each single variant of these elements would be present, on the average, in a concentration of about $10^{10}$ per milliliter. Mitchison at the National Institute for Medical Research and Nossal, Gordon L. Ada and their colleagues at the Walter and Eliza Hall Institute and the Australian National University, experimenting with low-zone tolerance, have shown that epitope concentrations ranging for different antigens from a million to $10^{12}$ epitopes per milliliter suffice either to suppress or to paralyze lymphocytes that can recognize the epitopes. Nisonoff and his co-workers at the University of Illinois College of Medicine and Humberto Cosenza and Heinz Köhler at the University of Chicago have shown that injecting into an animal antibodies against an idiotype suppresses lymphocytes that have receptors with idiotopes recognized by those antibodies. Leonard A. Herzenberg of the Stanford University Medical Center and Ethel Jacobson in our laboratory in Basel find that $T$ lymphocytes recognizing epitopes on the receptors of $B$ lymphocytes can suppress those $B$ lymphocytes.

What this adds up to is that lymphocytes are subject to continuous suppression by other lymphocytes and by antibody molecules with idiotopes or combining sites that fit. Some lymphocytes escape from suppression and divide. New lymphocytes emerge. Others remain suppressed or decay. The eigen-behavior is the dynamic steady state of the system as its elements interact. As the system expands in the course of development and later life, new idiotopes and new combining sites emerge. The "self"-epitopes of other tissues impinge on the network and cause certain elements to become more numerous and others less numerous. In this way each individual develops a different immune system.

Invading foreign antigens modulate the network; early imprints leave the deepest traces. A given foreign epitope will be recognized, with various degrees of precision, by the combining sites of a set of antibody molecules, and lymphocytes that are committed to producing antibody molecules of that set are then stimulated and become more numerous. That is not, however, the only imprint made by the foreign epitope. The set of combining sites that recognized the epitope also recognizes a set of idiotopes *within* the system, a set of idiotopes that constitutes the "internal image" of the foreign epitope. The lymphocytes representing the internal image will therefore be affected secondarily, and so forth in successive recognition waves throughout the network. The structural properties of the immune system and its eigen-behavior reside in these complex ramifications.

## Immune System and Nervous System

The immune system and the nervous system are unique among the organs of the body in their ability to respond adequately to an enormous variety of signals. Both systems display dichotomies: their cells can both receive and transmit signals, and the signals can be either excitatory or inhibitory. The two systems penetrate most other tissues of the body, but they seem to avoid each other: the "blood-brain barrier" prevents lymphocytes from coming into contact with nerve cells.

The nerve cells, or neurons, are in fixed positions in the brain, the spinal cord and the ganglia, and their long processes, the axons, connect them to form a network. The ability of the axon of one neuron to form synapses with the correct set of other neurons must require something akin to epitope recognition. Lymphocytes are 100 times more numerous than nerve cells and, unlike nerve cells, they move about freely. They too interact, however, either by direct encounters or through the antibody molecules they release. These elements can recognize as well as be recognized, and in so doing they too form a network. As in the case of the nervous system, the modulation of the network by foreign signals represents its adaptation to the outside world. Both systems thereby learn from experience and build up a memory, a memory that is sustained by reinforcement but cannot be transmitted to the next generation. These striking analogies in the expression of the two systems may result from similarities in the sets of genes that encode their structure and that control their development and function.

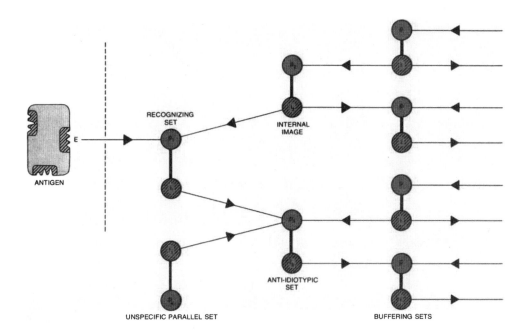

LYMPHOCYTE NETWORK is diagrammed in an effort to indicate how its steady-state ("eigen") behavior is established and how the network responds to an antigen. An epitope (E) on the antigen is recognized by a set ($p_1$) of combining sites on antibody molecules, both circulating antibody and cell-surface receptors. Cells with receptors of the recognizing set $p_1$ are potentially capable of responding to the antigenic stimulus (*arrowhead*) of epitope E, but there are constraints. The same molecules that carry combining sites $p_1$ carry a set of idiotopes ($i_1$). These are recognized within the system by a set of combining sites ($p_3$), called the anti-idiotypic set because they tend to suppress (*reverse arrowhead*) the cells of set $i_1$. (These idiotypes $i_1$ are also found on molecules with combining sites that do not belong to the recognizing set $p_1$ but rather are unspecific with regard to epitope E.) On the other hand, the set $p_1$ also recognizes internal epitopes $i_2$, which therefore constitute an internal image of the foreign epitope E. In the steady state, molecules of the internal image tend to stimulate cells of set $p_1$ and thus to balance the suppressive tendency of the anti-idiotypic set. When the foreign antigen enters the system, its stimulatory effect on recognizing set $p_1$ allows cells of that set to escape from suppression. (The same thing happens to unspecific cells of the parallel set $p_x$.) The resulting immune response to the antigen is modulated by the buffering effects of many more sets of combining sites and idiotopes (*right*), which have a controlling influence on the response.

## ABOUT THE AUTHOR

■ **Niels Kaj Jerne** (1911-1994) was awarded the Nobel Prize in Physiology or Medicine in 1984 along with Georges J. F. Köhler and César Milstein for theories concerning the specificity in development and control of the immune system and the discovery of the principle for production of monoclonal antibodies.

# The Molecular Biology of Poliovirus

*The agent of a once dreaded disease has become a tame laboratory organism, an excellent instrument for studying the multiplication of a virus and the cellular machinery it converts to that purpose.*

## Deborah H. Spector and David Baltimore

A virus is the most minimal of organisms, a protein-packaged snippet of genetic information that takes over the mechanisms of a living cell and converts them to a new purpose, which is simply the manufacture of new viruses. In the process viruses often kill cells and thus destroy tissue and cause disease. On the other hand, viruses are good laboratory animals, as it were, and even good laboratory instruments. Because they contain little genetic information and give rise to a limited number of recognizable products, they are ideal organisms in which to study the basic life processes: the replication of genetic information and its translation into protein. And because they reproduce only by appropriating particular elements of a cell's machinery, they are also excellent tools with which to probe into cells and learn how those cellular elements function.

One of the most effective viruses for these purposes is the once dreaded poliovirus. It has been intensively investigated since its discovery in 1909, originally of course because it attacks the nerve cells of the gray matter of the spinal cord and is the agent of poliomyelitis. It grows well and to high concentrations in laboratory cultures of the human tumor-cell line called HeLa. It is stable and easy to handle, and the development of the Salk and the Sabin vaccines in the 1950's rendered it safe. Poliovirus is also small even by viral standards; it seems to contain just enough genetic information to accomplish its own reproduction, with little extra genetic material to complicate matters. In the past 15 years a great deal has been learned about the multiplication of poliovirus and thus about certain cellular mechanisms. Here we shall give an account of some of those findings, after first describing the virus and its parasitic way of life.

An electron micrograph of the poliovirus virion (the virus particle) reveals a sphere 27 nanometers (millionths of a millimeter) in diameter.

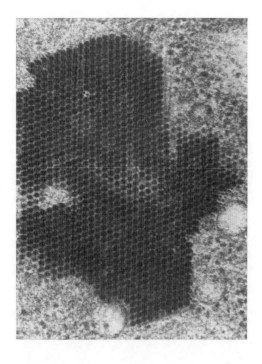

POLIOVIRUS PARTICLES lie closely packed in the cytoplasm of a cell. About 1,000 virions, or virus particles, form a crystalline array in the plane of the electron micrograph; other virions are out of the plane. The viruses are stained with lead and enlarged about 200,000 diameters in the micrograph, made by Samuel Dales of the Public Health Research Institute of the City of New York.

There are indications that the sphere is actually an icosahedron, a polyhedron with 20 faces, but that shape has not yet been resolved in micrographs. The virion consists only of protein and the nucleic acid RNA. The protein is formed into a capsid, a coat enclosing the nucleic acid that is constructed of 60 identical subunits arranged in icosahedral symmetry; it incorporates 60 copies each of four proteins that are designated VP 1, VP 2, VP 3 and VP 4.

Poliovirus RNA is a single chain of about 7,500 of the subunits called nucleotides, each of which consists of a ribose sugar component and one of four organic bases: adenine, uracil, guanine and cytosine. The nucleotides are linked by phosphate groups joining the carbon atom at position No. 3 on one sugar to the carbon at position No. 5 on the adjacent sugar. The two ends of the poliovirus RNA molecule are chemically distinct: at one end there is a free No. 3 position on the sugar and at the other end there is a No. 5 position with a terminal phosphate group. The two ends of the RNA molecule are therefore designated the 5' and the 3' ends. There are four kinds of nucleotide, named for their bases: adenylic acid, uridylic acid, guanylic acid and cytidylic acid, better known as *A, U, G* and C. For the most part they are distributed in what looks like random order, but their sequence is actually the code that specifies the genetic message. At the 3' end, however, there is a unique sequence: a string of about 75 adenylic acids in a row. Such a sequence is called polyadenylic acid, or poly-*A*. Its presence in poliovirus was discovered in 1972 by John A. Armstrong, Mary P. Edmonds, Hiroshi Nakazato, Bruce A. Phillips and Maurice H. Vaughan, Jr., of the University of Pittsburgh; later Yoshiaki Yogo and Eckard Wimmer, then at the Saint Louis University School of Medicine, showed that the poly-*A* constituted the 3' end of the molecule. Similar stretches of poly-*A* have been found on a number of cellular and viral RNA's.

In cells and in some viruses the genetic material is DNA, and RNA is the substance of various intermediary structures in the translation process; messenger RNA, for example, is transcribed from the original DNA and is thereupon translated into protein. In the poliovirus and certain other viruses that contain no DNA, the RNA is itself the carrier of the genetic information, which is translated into molecules of protein by the infected cell's machinery. Poliovirus RNA is both a genetic RNA and a messenger RNA: the information in the viral RNA is translated directly into protein, with no intermediate step of transcription into another RNA. The sequences of amino acids that make up the poliovirus proteins are specified by the 7,500 nucleotides of poliovirus RNA, and so the proteins must consist overall of about 2,500 amino acids (because it takes a code "word" of three nucleotide "letters" to specify one amino acid). That number of amino acids would make about 10 average-sized proteins. In a cell there are thousands of different proteins, and so poliovirus has only a tiny fraction of the information possessed by a cell. Since a virus is such a simple organism, however, and since it turns the cell's complex machinery to its own purposes, this small amount of information can be devastating: six hours after a single poliovirus attaches itself to a cell there are 100,000 new virions and the cell is dead.

The details of the events that accomplish that end—the replication of the viral RNA, its translation into proteins and the assembly of the proteins and the RNA into virions—are now starting to become clear in our laboratory at the Massachusetts Institute of Technology and in other laboratories. Beyond that, studies with poliovirus RNA and related RNA's have provided detailed information about the translation mechanisms in the eukaryotic (nucleated) cells of mammals and other organisms, and have at least narrowed down the possible functions of poly-A.

The infection of a cell in man or some other primate by poliovirus depends in the first instance on the virion's adsorption to the cell. There is an attraction between capsid protein and a cellular receptor so specific that even viruses closely related to the poliovirus bind to different receptors. (Naked infectious poliovirus RNA can infect cells that lack specific receptors, but in that case the proliferating virions are unable to bind to the cells, and so the infection is confined to a Single cycle.) The binding of the virion is followed by the release of RNA from its protein coat and its entry into the cytoplasm of the cell; just how the transfer is accomplished is still not known.

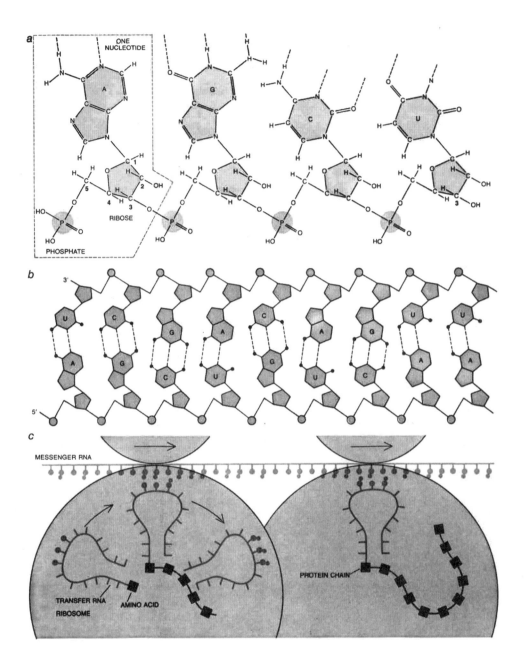

RNA, the genetic material of the poliovirus, is a chain of nucleotides (*a*), each consisting of a ribose sugar, a phosphate group and one of four organic bases: adenine (*A*), uracil (*U*), guanine (*G*) and cytosine (*C*). The sequence in which the nucleotides appear constitutes the genetic code. The phosphates join the carbon atom at position No. 3 on one sugar to carbon No. 5 on the next sugar; an RNA molecule therefore has distinct 3' and 5' ends. RNA replicates (forms copies of itself) through assembly of complementary nucleotides to form a complementary strand (*b*) according to base-pairing rules: *A* pairs with *U* and *G* pairs with *C*. Hydrogen bonds (*broken lines*) link complementary bases. RNA, in the form known as messenger RNA, is the medium whereby the genetic code is translated into protein (*c*) in the cellular structures called ribosomes. The RNA is "read" by small transfer RNA's that recognize three-letter codons (sequences of three bases) as codes for specific amino acids. The elongating chain of amino acids folds to become a protein.

Poliovirus replicates in the cytoplasm of the infected cell, not in the nucleus; even cells whose nucleus has been removed will grow the virus. The time course of viral multiplication varies considerably, depending on such factors as the specific virus strain, the host cell, the nutritional state of the cell and the multiplicity of infection. When HeLa cells are infected under optimal conditions, no new virus is detectable until about three hours after infection; during the intervening "eclipse" period the viral functions are getting under way. During this phase the virus also inhibits the synthesis of cellular RNA and protein, thus freeing most of the cell's machinery for the virus's purposes.

As we have indicated, the injected poliovirus RNA must serve two basic functions in order to initiate an infection. First, it must act as the messenger RNA to be translated into viral protein. More specifically, if the viral RNA is to be replicated, there must be an active replicase (a replicating enzyme, in this case RNA-dependent RNA polymerase). Since there is little evidence of replicase activity in uninfected cells, it appears that the synthesis of at least some of the enzyme is directed by the viral RNA. Second, the viral RNA must act as a template for the synthesis by the replicase of a new molecule of RNA, one that has a sequence of nucleotides that is the mirror image of the viral-RNA sequence. This is accomplished through the fundamental complementarity in the structure of nucleic acids: according to the base-pairing rules, A always pairs with U and G always pairs with C. The virion RNA is called plus RNA and the complementary RNA that is synthesized in the cell is called minus RNA. The minus RNA serves in turn as a template for the synthesis of new copies of plus RNA.

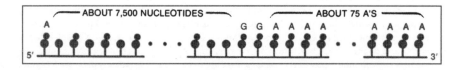

POLIOVIRUS RNA is a single chain of nucleotides. There is an A at the 5' end, followed by roughly 7,500 nucleotides that code for the viral proteins. Then, after two G's, there is a distinctive sequence of about 75 A's, or what is called poly-A, at the 3' end of the strand.

L et us follow the translation process step by step. The viral-RNA molecule becomes attached to cellular ribosomes, the major constituents
of any cell's protein-synthesizing system. The RNA and several ribosomes
form a polyribosome, the structure on which protein synthesis is carried
out. Here the viral RNA acts as messenger RNA to specify the amino acid
sequence of the various viral proteins. The first viral RNA to enter the cell
has to compete in the translation process with cellular RNA. Subsequent
molecules of viral RNA have an easier time because the virus somehow
interferes progressively with the ability of the host cell's ribosomes and
messenger RNA to interact and thus makes the ribosomes available for
synthesizing viral proteins. HeLa-cell polyribosomes, for example, begin
to disintegrate immediately after poliovirus infection; they are replaced
by larger polyribosomes, and it is in these larger structures that the synthesis of viral proteins proceeds. One of the most interesting unanswered
questions of virology is how poliovirus and many other viruses selectively
inhibit host-cell protein synthesis and take over the ribosomes to make
viral proteins.

Many nucleic acid molecules appear to have "punctuation points" along
their length that signify the beginning and end of a gene, which can be

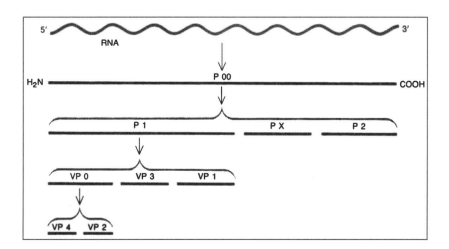

TRANSLATION of poliovirus RNA results in the synthesis of a single very large protein designated P00,
which is immediately cleaved to yield three proteins: P 1, PX and P 2. P X and P 2 are probably forms of
replicase, the enzyme that polymerizes new RNA strands. P 1 is the precursor of structural proteins that
form the coat of new virus particles. P 1 is cleaved to form VP 0, VP 3 and VP 1; finally VP 0 is cleaved
into VP 4 and VP 2.

defined as a sequence of nucleotides that codes for a single kind of protein. The poliovirus RNA molecule appears to have only two punctuation points: one "start" signal and one "stop" signal. Ribosomes therefore attach themselves near the 5' end of the RNA and proceed along the entire length of the molecule, forming a giant chain of amino acids with a molecular weight of about 250,000. This huge protein, really a polyprotein, is then systematically cleaved by proteolytic enzymes. (The early stages of the cleavage process are very fast; we can only identify the complete polyprotein by incorporating into it chemical analogues of some of the normal amino acids, which prevents cleavage.) The polyprotein, which we call P 00, is cleaved at two sites to make three products, P 1, P X and P 2. Probably either P X or P 2 represents the replicase enzyme; perhaps both of them do.

P 1 is the precursor molecule for the viral capsid proteins, which are successively cut from it and aggregated into structures of increasing size. First P 1 is cleaved to yield three direct products, VP 0, VP 1 and VP 3, that seem to remain aggregated as a single structural subunit. Five of these subunits aggregate into a fivefold molecule and then 12 of these pentamers come together to form an empty shell, the procapsid. A procapsid combines with a molecule of viral RNA, making a provirion, Finally the VP 0 molecules of the provirion are cleaved into two distinct proteins, VP 2 and VP 4. We have distinguished each of these steps in the laboratory by separating the different structures and particles on the basis of their sedimentation rates and then analyzing their protein and RNA content.

Whereas most poliovirus research has been conducted with HeLa cells, in the past four years we and our associates Lydia Villa-Kamaroff and Harvey F. Lodish have developed systems that make it possible to investigate the translation of poliovirus RNA outside the cell. When we add poliovirus to extracts of various cells (including HeLa cells) under the proper conditions, we can detect translation, at a low but measurable efficiency, into a protein that is recognizable as P 00, the largest poliovirus protein. The synthesis is initiated at only one site on the RNA molecule, which is an indication of correct translation in a cell-free system; the protein chains do tend, however, to stop growing before they are completed. We are now exploiting these cell-free systems to study various aspects of the transla-

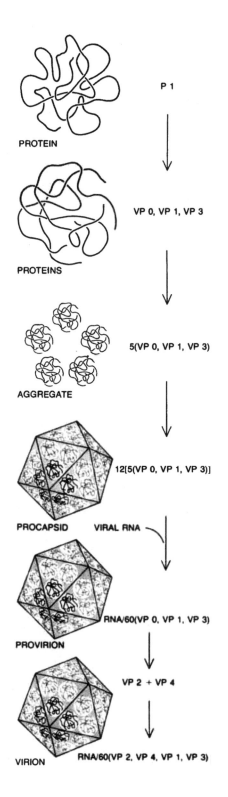

P 1

PROTEIN

VP 0, VP 1, VP 3

PROTEINS

5(VP 0, VP 1, VP 3)

AGGREGATE

12[5(VP 0, VP 1, VP 3)]

PROCAPSID          VIRAL RNA

RNA/60(VP 0, VP 1, VP 3)

PROVIRION

VP 2 + VP 4

VIRION          RNA/60(VP 2, VP 4, VP 1, VP 3)

CAPSID IS FORMED in a series of steps in which the capsid proteins are cleaved and aggregated. A likely pathway (details are still being studied) is shown here. P 1 is cut into three smaller molecules that remain assembled in a structural subunit, five of which aggregate. Twelve aggregates are assembled (possibly as shown here, at the vertexes of an icosahedron) into the empty procapsid. The provirion encloses a viral. RNA molecule. Finally the protein VP 0 is cleaved and the new virion is complete.

tion process in greater detail and also to try to learn how the virus inhibits cellular protein synthesis.

Once some of the infecting RNA molecules have been translated, the enzyme replicase becomes available for the transcription of complementary RNA (the minus strand) from viral RNA (the plus strand). Presumably the replicase attaches itself to the poly-A tail at the 3' end of the viral RNA and proceeds to synthesize a minus strand, which begins with a sequence of complementary poly-U at its 5' end. As each minus strand is completed it serves as a template for the simultaneous transcription of several new viral (plus) molecules. We have isolated a structure reflecting that stage: several plus strands of graduated lengths partially bonded to the minus strand on which they are being synthesized; we call such a structure the replicative intermediate. (We still do not know whether two different replicases are required for transcription from plus to minus and from minus to plus or whether a single enzyme performs both functions.) Each replicative intermediate transcribes plus strands for only a short time. Apparently a plus strand then fails to peel off and remains bonded to its minus strand, so that what was a functioning replicative intermediate becomes a double-strand RNA molecule. Such double-strand molecules accumulate until at the end of the infection cycle they account for a fairly large proportion of the total viral RNA in the infected cell.

Experiments in our laboratory and in that of Wimmer, who is now at the State University of New York at Stony Brook, suggest that the poly-A tail at the 3' end of viral-RNA molecules is transcribed from the poly-U sequence at the 5' end of the minus strand. The poly-U sequence is heterogeneous, ranging in length from about 50 to more than 200 nucleotides, whereas the poly-A sequence runs about 75 nucleotides. It is possible that a plus strand can be pushed off its minus-strand template before all the U's are transcribed by the arrival of the next replicase molecule with its plus strand. We have noted that late in infection, when the rate of RNA synthesis has slowed down and there may be less such pushing, there are poly-A tails with more than 75 A's on many newly synthesized viral-RNA molecules, suggesting that the length of the tail is determined by the period during which the viral-RNA molecule remains associated with the replicative intermediate.

The synthesis of poliovirus RNA can be divided into two periods. During the eclipse phase (the first three hours after infection) RNA synthesis proceeds at an exponential rate; toward the end of that period protein synthesis reaches a peak. Then for about an hour the rate of synthesis remains constant and viral RNA accumulates at a linear rate. The cause of the switch from exponential to linear synthesis is not known, but it comes just about when the cell is beginning to manufacture new virions from the newly synthesized RNA and proteins. At the end of the hour-long second phase the rate of RNA synthesis falls off rapidly, as does the synthesis of viral protein; the cell is by now functionally dead, although some viral activity may continue for a while before the cell bursts. The changing events of the viral multiplication cycle, such as the phases of RNA synthesis, appear not to be the result of direct temporal control of various genes. In some other viruses particular genes are turned on and off at different times; there are "early" enzymes, for example, that are synthesized at the direction of genes activated early in the infective process. In the case of poliovirus the same set of viral-gene products is synthesized throughout the infection, albeit at different rates; something other than genetic control of protein synthesis must regulate the changes in rate.

To take one example, every newly synthesized RNA molecule has three possible destinies. It can serve as a template for the transcription of a minus strand, it can serve as a template for translation into protein or it can become associated with capsid proteins and form a new virion. We know that as the infection proceeds more and more of the RNA goes to form virions, presumably because more capsid protein is available for the RNA to associate with. For a time it appeared that the poly-A sequence might be involved in regulating the destiny of viral RNA, but that has not proved to be the case.

Does the poly-A have any important role in the viral growth cycle, then? That is what we have been trying to determine for the past year. We began by analyzing the poly-A of RNA molecules involved in replication, translation and the formation of virions; there was no difference in the average size of the poly-A tail in the three conditions. The only consistent difference in size we could find was the temporal one to which we have alluded: viral-RNA molecules with longer than normal tails are produced late in the infection cycle. These long-tailed molecules are never found in virions; perhaps it is the excess poly-A that keeps them out. That still would not explain the normal role of normal amounts of poly-A, however.

In order to learn whether the poly-A mattered at all to the virus we asked what would happen to the infectivity of naked poliovirus RNA if it lacked the poly-A tail. The poly-A could be removed by an enzyme, ribonuclease H, that has a special property: it breaks down only that part of a single-strand RNA molecule which is bonded to a complementary strand of DNA. In DNA thymidylic acid (T) takes the place of the U in RNA and therefore bonds to A. By hybridizing a DNA molecule consisting of poly-T to the poly-A of poliovirus RNA molecules we were able to digest away more than 80 percent of the poly-A with ribonuclease H. The resulting RNA, now deficient in poly-A, turned out to be less than 5 percent as infectious as poliovirus RNA with a normal complement of poly-A. That could be because the RNA needs poly-A in order to get into the cell. We cannot test that possibility, however, because one can detect RNA entry only through its infectious effects. In any case it seems unlikely. We have assumed, therefore, that the poly-A does not prevent the entry of RNA but rather plays a critical role in some intracellular step in the multiplication of poliovirus.

When we examine cells exposed to RNA that is deficient in poly-A, we find no evidence that the RNA can multiply. In view of the RNA's double function in multiplication, as a template for protein synthesis and as a template for its own replication, the poly-A might be necessary either for translation or for replication. One experiment argues against a role for poliovirus poly-A in the translation of viral proteins: at least in two cell-free protein-synthesizing systems, poliovirus RNA that is deficient in poly-A is translated as efficiently as RNA with a normal complement of poly-A. That leaves us with the likelihood that poly-A plays some necessary role in the replication of poliovirus RNA, but the nature of that role is obscure.

**S**equences of poly-*A* are not peculiar to poliovirus RNA. They are also found as integral elements at the 3' ends of a number of other RNA's, notably the messenger RNA of nucleated cells. In spite of investigations by a number of laboratories the biological function of these sequences is still not known.

The mechanism of messenger-RNA synthesis in nucleated cells is still an unsettled question. One widely accepted hypothesis is that the messenger RNA is derived from longer molecules called heterogeneous nuclear RNA, a collection of RNA molecules of different lengths that are transcribed from the cell's DNA and are found in the nucleus. A sequence of about 200 nucleotides of poly-*A* is somehow added to the 3' end of some of the heterogeneous RNA molecules. This posttranscriptional addition of poly-*A* differs from the transcriptional addition we have described for poliovirus RNA, which utilizes poly-*U* as a template. For one thing, in cellular DNA there are no long sequences of poly-*T*, the DNA analogue of RNA's poly-*U*, that could serve as templates for poly-*A* transcription. Furthermore, there is a lapse of about 10 minutes between the transcription of the heterogeneous nuclear RNA molecules and the addition of poly-*A* to some of them. It appears that pieces of the heterogeneous RNA that have poly-*A* tails are cut down to the size of messenger RNA and then transported through the nuclear membrane into the cytoplasm of the cell, there to become messenger-RNA molecules that attach themselves to ribosomes and are translated into protein. While the messenger-RNA molecules are in the cytoplasm their poly-*A* tails gradually become shorter, until they are only about 100 nucleotides long. (On the other hand, there is evidence that poly-*A* is sometimes added to messenger RNA in the cytoplasm.)

Several enzymes that act to add poly-*A* to RNA have been found in both the nucleus and the cytoplasm of many kinds of cells. The ubiquity of these enzymes implies not only that the addition of poly-*A* could take place in either the nucleus or the cytoplasm but also that the poly-*A* plays some significant role in the genesis or functioning of messenger RNA. The role is presumably not in replication (as it apparently is in the case of poliovirus RNA) since no evidence has been found for the replication of cellular RNA. And the role is presumably not in the transcription of RNA from

DNA because most of the poly-A is added some time after transcription.

It has been suggested that the primary function of poly-A has to do with the processing of messenger RNA and its transport from the nucleus to the cytoplasm. This hypothesis seems to be negated by the finding that there are perfectly good messenger RNA's that are made in the nucleus of cells and transported to the cytoplasm and that do not contain poly-A.

It seems more likely that poly-A is involved in the translation of messenger RNA into protein. Here the evidence appears somewhat contradictory. There are some indications, first of all, that poly-A is not absolutely necessary for translation. That was the implication of our findings with poliovirus RNA in cell-free systems. Similarly, the efficiency of translation of rabbit messenger RNA coding for hemoglobin or of mixed cellular messenger RNA has been shown to be only slightly lower than normal when the RNA is deficient in poly-A, again in cell-free systems.

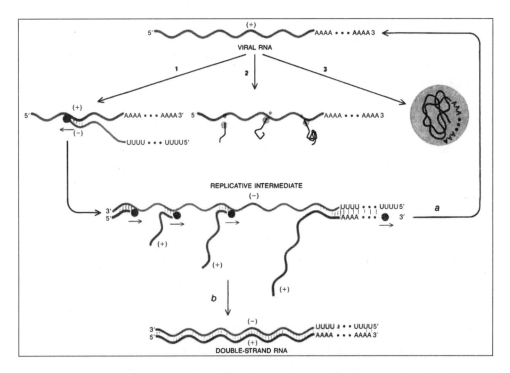

NEWLY SYNTHESIZED VIRAL RNA can go one of three ways. It can be transcribed into a minus strand (1), be translated into protein (2) or be associated with capsid proteins to form a new virion (3). If it is transcribed into a minus strand, the minus strand serves as a template for the transcription of more viral RNA that either is recycled (0) or ends up in a double strand of RNA (b).

Protein synthesis is less efficient in most cell-free systems, however, than it is in cells. When all the cellular machinery is present, the results are somewhat different. Investigators at the Free University of Brussels and at the Weizmann Institute of Science in Israel recently injected rabbit-hemoglobin messenger RNA with and without poly-A tails into the egg cells of the toad *Xeno-pus laevis*. (The egg's protein-synthesizing system had previously been shown to be capable of translating most messenger RNA from other species.) During the first hour after infection the two kinds of RNA supported identical rates of hemoglobin synthesis. After that, however, the rate of synthesis began to decline in the cells injected with RNA that lacked poly-A, and by the fifth hour it was only half the original rate. Translation of the RNA that contained poly-A, on the other hand, continued at the original rate for at least 48 hours. Although it is difficult to formulate any general conclusion based on the translation of a specific RNA in a very specialized cell, the result suggests that poly-A may help to stabilize cell messenger RNA's in some way and thus keep them operating at maximum efficiency.

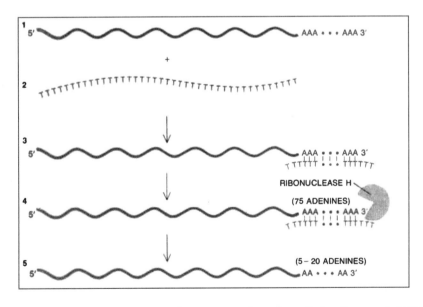

POLY-A is removed from poliovirus RNA with an enzyme, ribonuclease *H*, that digests only RNA bonded to complementary DNA. Viral RNA (1) is combined with poly-*T*, the DNA analogue of poly-*U* (2). The poly-*T* bonds to poly-*A* (3), and the ribonuclease *H* degrades most *A* nucleotides (4), leaving a viral-RNA molecule deficient in poly-*A* (5).

The poly-A story, incomplete as it is, is typical of the kinds of knowledge that emerge from the study of viruses and the cells they infect. By studying polio-virus multiplication we have learned some things about HeLa cells. The work has defined specific receptors on the cell surface, has explained some elements of the cellular machinery for protein synthesis and has shown how newly made proteins are processed. In the case of poly-A we have been able to demonstrate a role in viral metabolism for a structure that is ubiquitous in cells but has no clearly defined role in the cell. In the future we can expect that more detailed knowledge of how a number of mammalian viruses grow will help to unravel the complex mechanisms that enable cells to live and fulfill their varied functions.

## ABOUT THE AUTHOR

**David Baltimore** (born 1938) was awarded the Nobel Prize in Physiology or Medicine in 1975 along with Howard M. Temin and Renato Dulbecco for their discoveries concerning the interaction between tumour viruses and the genetic material of the cell. He is currently the Robert A. Millikan Professor of Biology at the California Institute of Technology.

# Oncogenes

*They are genes that cause cancer.*
*They were first found in viruses, but their*
*evolutionary history implies that normal vertebrate*
*cells have genes whose abnormal expression*
*can lead to cancerous growth.*

## J. Michael Bishop

C an the cancer cell be understood? Since no one can yet explain how a normal cell controls its growth, it may seem foolhardy to think the abnormal rules governing the growth of a cancer cell can be deciphered. Yet the history of biology records many instances in which the study of abnormalities has illuminated normal life processes. Recent developments in cancer research have added a dramatic new example. For the first time investigators have perceived the dim outline of events that can induce cancerous growth. Enzymes that catalyze those events have been identified, and so have the genes specifying the structure of the enzymes.

These advances have come from the study of viruses that induce tumors. Recent years have seen an enthusiastic search for viruses that cause cancer in human beings. The search has been largely unsuccessful, leading many

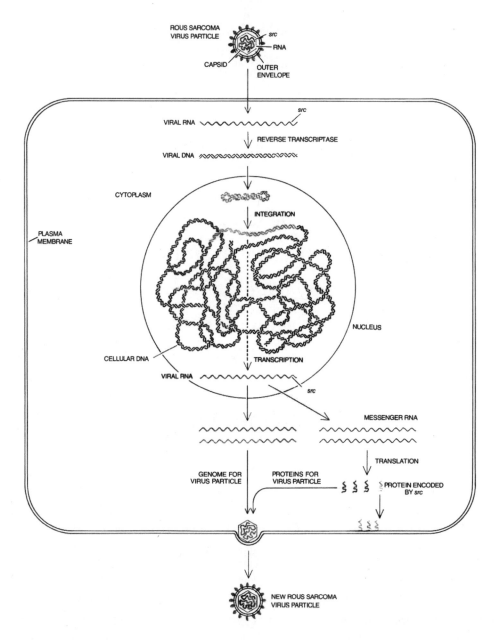

ONCOGENE called *src*, which induces a cancer (a sarcoma) in chickens, is part of the small RNA genome of the Rous sarcoma virus, a "retrovirus." When a retrovirus infects a cell, the viral RNA is copied into double-strand DNA by the enzyme reverse transcriptase, which is supplied by the virus. The DNA forms a circle and then becomes integrated into the host cell's DNA. When the host DNA is transcribed into RNA by cellular enzymes (not shown), the viral DNA is also transcribed (*broken arrow*). Some of the viral RNA provides copies of the viral genome for inclusion in new virus particles and some of it is processed to make messenger RNA, which is translated by the cell's protein-synthesizing machinery into viral proteins. Several of these proteins are incorporated into new virus particles. The product of the *src* gene, however, is not a component of the virus particle. It is an enzyme called a protein kinase, which binds to the inner surface of the cell's plasma membrane and phosphorylates cellular proteins, thereby converting the cell to cancerous growth.

informed observers to doubt that viruses will ever prove to be a major cause of human cancer. Some viruses do induce tumors in other animals, however, and investigators have been studying these tumor viruses, attempting to define fundamental derangements of the cell that are responsible for cancerous growth. That quest has struck gold.

Although the genes implicated in the development of cancer were first observed in work with viruses, they are not native to the viruses. Indeed, it has turned out that the genes are not even peculiar to cancer cells. They are present and functioning in normal cells as well, and they may be as necessary for the life of the normal cell as they appear to be for the unrestrained growth of a cancer. A final common pathway by which all tumors arise may be part of the genetic dowry of every living cell.

## Tumor Viruses

A virus is little more than a packet of genetic information encased in a protein coat. The information can be embodied in either DNA or RNA (whereas in the cells of higher organisms the genetic archive invariably consists of DNA). Both DNA and RNA are long strands of four of the chemical units called nucleotides. The sequence of the nucleotides constitutes a coded message, punctuated into the discrete units called genes. The instructions encoded in genes are carried out in various ways. Most commonly the sequence of nucleotides specifies the order in which amino acids are assembled to form a particular protein, typically an enzyme or a structural element. Viruses can have fewer than five genes and never have more than several hundred, whereas the cells of more complex organisms have a genome, or total genetic complement, of tens of thousands of genes. The reproduction of viruses mimics the processes by which cells grow and divide, but the simplicity of viruses makes them much easier than cells to study and understand.

In cells DNA is transcribed into a strand of messenger RNA and the RNA is translated into protein. An infecting virus insinuates its genetic information into the cellular machinery, so that the cell synthesizes viral proteins specified by viral genes. The proteins synthesize many copies of the viral genome, construct new virus particles and execute any other instructions of the viral genes. In some instances the instructions include a command that converts the host cell to a cancerous state.

The existence of tumor viruses was first suspected at the turn of the century. A critical discovery came in 1910, when Peyton Rous of the Rockefeller Institute for Medical Research showed that a cell-free filtrate from chicken tumors called sarcomas could induce new sarcomas in chickens. His reports were not well received; eventually Rous abandoned his work on tumor viruses because of his peers' disapproval. Decades later the reality of the virus first identified by Rous, and of other tumor viruses as well, was established beyond doubt by purification with physical techniques and visualization with the electron microscope. Tumor viruses became workaday agents in cancer research. In 1966, at the age of 85, Rous was awarded a Nobel prize.

Some tumor viruses are oncogenic (that is, they induce tumors) only in animals that are not their host in nature, whereas other tumor viruses are oncogenic in their natural host. Such differences are understood only in part, but for the investigator they are of no great concern. The ability to induce tumors at will with a rather simple and well-defined agent has been a great boon to cancer research, even if it is sometimes necessary to resort to an unnatural combination of virus and host.

## Transformation

**M**any tumor viruses have a particularly valuable property: they elicit cancerous changes in cells in an artificial culture medium. This "transformation" of cultured cells makes it possible to examine the interaction of a tumor virus with a host cell under controlled conditions and to avoid the difficulties associated with experiments in animals. It is important to remember, however, that some tumor viruses do not transform cells in culture and yet are powerful oncogenic agents in animals.

The ability or inability of a virus to transform cultured cells is connected with its mechanism of oncogenesis. Two patterns have been recognized. Some viruses have a single gene that is solely responsible for their capacity to induce tumors, or in some cases a few such "oncogenes." The action of viral oncogenes is rapid, and it predominates over the activity of all other genes in the cell. Most viruses with oncogenes (and perhaps all of them) can transform cells in culture; the capacity for transformation is provisional evidence that a virus has an oncogene. Other viruses lack oncogenes and induce tumors by more subtle means. Such tumor viruses act slowly in animals, in many cases taking from six to 12 months to give rise to a tumor, in contrast to the few days or weeks required by an oncogene virus. And they do not transform cells in culture.

Both forms of oncogenesis are characterized by the persistence of the viral genome in the host cell for as long as the cell survives. In most instances viral ONA has been integrated, or chemically stitched, into the DNA of the host cell, but the genome of some tumor viruses appears to survive within the cell as a separate unit and to reproduce independently. At the moment it seems that the persistence of the viral genome is necessary for viral oncogenesis, either to maintain the influence of an oncogene over the cell or to sustain the less direct effects of viruses that induce tumors but do not carry an oncogene. The mysteries of viral oncogenesis have occasionally prompted the hypothesis of a "hit and run" mechanism in which a transient virus infection triggers a sequence of events eventuating in a tumor, with no trace of the virus necessarily persisting in the tumor cells. There is now very little evidence to support such models.

## Retroviruses

The sarcoma virus discovered by Rous belongs to a family known as the retroviruses, which are the only tumor viruses with an RNA genome. Retroviruses have provided the most coherent view of oncogenesis now available. Three features of retroviruses account for their utility in the analysis of tumor development. First, they have been found in a large number of vertebrate species and they induce many types of tumors: experimental models for most major forms of human cancer. Second, it is relatively easy to identify and isolate retrovirus oncogenes and to find their products, and so they have provided the first glimpse of the chemical processes responsible for cancerous growth. Third, retrovirus oncogenes do not appear to be indigenous components of the viral genome; instead they seem to have been copied from genes of the vertebrate host in which the virus replicates. There is reason to suspect that the cellular genes from which the retrovirus oncogenes apparently arose are themselves involved in the production of tumors induced by agents other than viruses. Thus tumor virologists engaged in the arcane endeavor of tracking the evolutionary origin of oncogenes have been led to genetic mechanisms that may underlie many forms of cancer.

Retroviruses derive their name from a feature of their life cycle that makes them unique in biology: their RNA must be transcribed "backward" into DNA for them to propagate. This unusual process is accomplished by an enzyme called reverse transcriptase. The enzyme was discovered in the particles of viruses such as the Rous sarcoma virus in 1970 by David Baltimore of the Massachusetts Institute of Technology and by Satoshi Mizutani and Howard M. Temin of the University of Wisconsin. The discovery was important on several counts. It scuttled the widely held misconception that genetic information could flow only from DNA to RNA. It triggered a surge of research on retroviruses by clarifying the previously obscure mechanism of their replication. And it provided an essential reagent for the developing technology of genetic engineering with recombinant DNA.

The life cycle of a retrovirus is a marvel of cooperation between parasite and host. The success of virus infection depends on the lavish hospitality

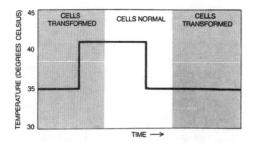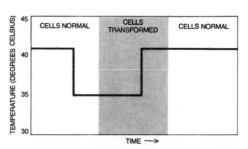

CONDITIONAL MUTATION of a gene results in the reversible inactivation of the gene or the enzyme it encodes. In the case of a temperature-sensitive mutant of *src*, infected cells are transformed when they are cultured at 35 degrees Celsius but revert to (or remain in) the normal state at 41 degrees. The discovery of such mutants implies that a viral gene must be responsible for transformation, that the action of the gene is probably mediated by a protein product and that the sustained action of the gene is necessary for transformation.

offered by the cell, and yet the virus retains much authority to control events. During the early hours of infection the viral RNA genome is transcribed into DNA by reverse transcriptase. The viral DNA is then integrated into the cell's genome, with the result that viral genes are replicated along with cellular genes and are expressed by the machinery of the cell.

In many cases a retrovirus infection is innocuous to the cell. The virus acquires a new and potentially enduring home; new virus particles are manufactured and leave the cell, and yet the cell suffers no damage. The partnership can go awry, however, as a result of either of the two kinds of viral oncogenesis mentioned above. If the virus carries an oncogene, the activity of the gene can convert the cell to cancerous growth. If the virus lacks an oncogene, the integration of the viral DNA can interfere with a cellular gene at or near the point of insertion; in other words, the insertion can cause a mutation in the host cell's genome. Mutations at certain sites may engender cancerous growth. The induction of tumors by oncogenes and induction by the consequences of integration appear at first to be quite dissimilar events, but I shall show below that they are intimately related.

## The *src* Gene

The oncogene of the Rous sarcoma virus was the first to yield to experimental analysis. An important early step was taken in 1970, when G. Steven Martin of the University of California at Berkeley identified temperature-sensitive "conditional" mutations that affect the ability of the virus to transform cells in culture. A conditional mutation is a powerful tool because it makes possible the reversible inactivation of a gene. When cultured cells infected with temperature-sensitive Rous sarcoma viruses are maintained at a "permissive" temperature, they are transformed. When the temperature is shifted to a higher, "restrictive" one, within hours the cells regain a normal appearance, only to be transformed once more when the temperature is again lowered. The interpretation is that at a restrictive temperature a mutated gene is inactivated. Transformation, then, is due to the action of a gene, which must be expressed continuously to maintain the cancerous state. (In most cases the elevated temperature probably does not act directly on the gene itself. Instead the mutation alters the structure of the protein product of the gene, with the result that the activity of the protein is impaired by the restrictive temperature.)

The gene first glimpsed by Martin is now called *src* (for sarcoma, the tumor it induces); it is the oncogene of the Rous sarcoma virus. The *src* gene was soon made more tangible by Peter H. Dues-berg of Berkeley and by Charles Weissmann, Martin Billeter and John M. Coffin of the University of Zurich. They worked with strains of the Rous sarcoma virus that had been isolated by Peter K. Yogt of the University of Southern California. The strains are "deletion" mutants that have lost the oncogene and are therefore incapable of inducing tumors or transforming cells in culture. Duesberg and Weissmann and his colleagues fragmented the genomes of deletion mutants and of wild-type (oncogenic) viruses with the enzyme ribonuclease. By determining which fragment was missing in the mutants they were able to identify the oncogene as a segment of RNA near one end of the Rous sarcoma virus genome.

In the past few years the powerful new techniques of genetic engineering have been exploited to define oncogenes more precisely and to test their cancerous potential. DNA can now be cut into fragments at specific sites with the aid of a battery of enzymes called restriction endonucleases. Particular fragments can be grown in quantity in bacteria, then reisolated and inserted into cultured cells, where the genes carried by the DNA are expressed. In this way one can cut viral DNA into pieces that each carry a single gene and learn which of the pieces cause transformation. Analysis of the DNA of the Rous sarcoma virus has revealed a single gene capable of transforming cells; the gene encodes a single protein product. The implication is that one gene, by directing the synthesis of one protein, can bring about the changes characterizing a cancer cell. To know that protein and how it acts is to have in view the events that can generate a malignant tumor.

The protein encoded by *src* is known, owing largely to the work of Raymond L. Erikson of the University of Colorado School of Medicine and his colleagues. They began by identifying a protein that is synthesized in the test tube under the instructions of the wild-type Rous sarcoma virus

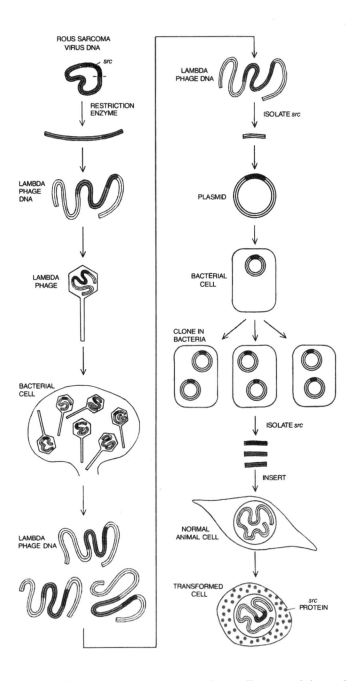

VIRAL ONCOGENE is purified, and its capacity to transform cells is tested, by methods of genetic engineering. The circular DNA of the Rous sarcoma virus is isolated from newly infected cells, cleaved with a restriction enzyme and inserted into the DNA of phage lambda, a bacterial virus. The growth of the phage in bacteria makes large quantities of the viral DNA, which is cleaved with a restriction enzyme to yield a fragment bearing only *src* and a bit of flanking DNA. The fragment is inserted into a plasmid (a small circle of bacterial DNA), which is introduced into bacteria for further amplification. Now many copies of the *src* fragment can be excised from the plasmids, purified and introduced into animal cells in culture. Fragment directs synthesis of a viral protein that induces transformation. Experiment was done by William J. DeLorbe and Paul A. Luciw to show that *src*, acting alone, gives rise to cancerous growth.

genome but not under the instructions of the genome of a deletion mutant lacking *src*. Then they raised rabbit antibodies to a putative *src* protein by inducing tumors in rabbits with the Rous sarcoma virus. The antibodies combined specifically with the protein synthesized in the test tube and

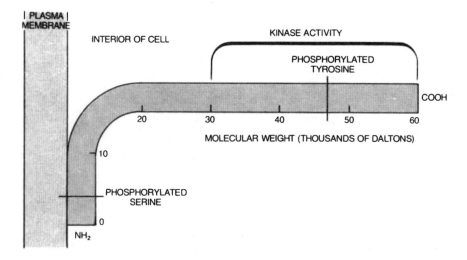

ONCOGENE PRODUCT, a protein designated pp60v-*src*, is a chain of about 520 amino acid subunits. It is a protein kinase: an enzyme that adds phosphate groups to proteins. The protein is bound to the plasma membrane of cells by a domain near the amino (NH2) end; the phosphorylating site is in the other half of molecule. Enzyme itself is phosphorylated at two sites.

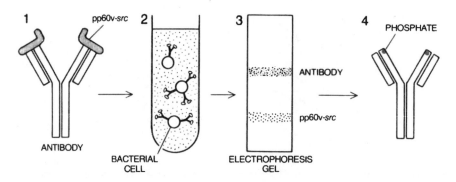

PROTEIN-KINASE ACTIVITY of pp60v-*src* was discovered fortuitously. Rabbits in which tumors have been induced by the Rous sarcoma virus make antibodies to pp60v-*src*, and the antibodies form complexes with the protein (1). The complexes were immobilized on the surface of bacteria and a source of radioactively labeled phosphate (*dots*) was added (2). When pp60v-*src* and the antibody were separated by gel electrophoresis, the antibody band on the gel showed radioactivity (3): the immobilized pp60v-*src* had catalyzed the transfer of the phosphate to the amino acid tyrosine at a single site in each of the antibody's two heavy chains (4).

also with an identical protein in cells transformed by *src*. These findings persuasively identified a protein that is encoded by *src* and is responsible for the effects of the gene. The protein was designated pp60v-*src;* the "pp" signifies that it is a phosphoprotein (a protein to which phosphate groups are attached), the "60" refers to its molecular weight of 60,000 daltons and "v-*src*" indicates its genetic origin is the viral gene *src*.

## A Cancer Enzyme

How does the protein product of the *src* gene convert a cell to cancerous growth? That seemed to be a daunting question when the protein was isolated. Yet a first answer came quickly when it was discovered that pp60v-*src* is a protein kinase: an enzyme that attaches phosphate ions to the amino acid components of proteins in the reaction known as phosphorylation. The discovery was made by Erikson and his colleague Mark S. Collett and, independently, by Arthur Levinson, working with Harold E. Varmus and me in our laboratory at the University of California School of Medicine in San Francisco.

Soon thereafter Tony Hunter and Bartholomew M. Sefton of the Salk Institute for Biological Studies reported that pp60v-*src* attaches phosphate ions specifically to the amino acid tyrosine. That put pp60v-*src* outside the known classes of protein kinases, which phosphorylate the amino acids serine and threonine. The phosphorylation of tyrosine has turned out to be a common characteristic of oncogene-encoded enzymes; surprisingly, it also has a role in regulating the growth of normal cells.

Not many years ago phosphate appeared to most biologists to be a mundane material and its transfer to proteins a humble event. Now it is clear that the phosphorylation of proteins is one of the central means by which the activities of growing cells are governed. One enzyme, by phosphorylating a number of proteins, can vastly alter the functioning of a cell. In the case of pp60v-*src* two modes of action have been proposed. The enzyme could phosphorylate a single protein, precipitating a cascade of events that together give rise to the properties of a cancer cell; alternatively, the enzyme could phosphorylate numerous proteins, directly affecting the functions of each of them and perhaps precipitating secondary events or even cascades in turn. What little is known at the moment makes it seem likely that the second alternative correctly describes the action of pp60v-*src*.

Can the phosphorylation of tyrosine subunits in cellular proteins account for the ability of *src* to induce tumors? Hunter and his colleagues have shown that the amount of phosphorylated tyrosine in a cell increases approximately tenfold as a result of transformation by *src*. The increase is regarded as a manifestation of the activity of pp60v-*src*. The critical questions now are: What cellular proteins are phosphorylated by the enzyme and what are their functions? There are only a few clues, none of which can yet account for the unrestrained growth of the tumors induced by *src*. The pursuit of targets for pp60v-*src* is under way in many laboratories.

## Site of Action

One approach is to find out where in the cell pp60v-*src* acts, in the hope of learning what proteins it affects and what those proteins do. Early studies indicated that the products of viral oncogenes might take up residence in the nucleus of the cell, where they could meddle directly with the apparatus responsible for replicating the cellular DNA and so drive the cell to unrestrained growth. Experiments by Hartmut Beug and Thomas Graf of the Max Planck Institute for Virus Research in Tübingen showed, however, that the effects of the *src* protein can be detected even in cells from which the nucleus has been removed. It came as no surprise, then, when several workers found that little if any of the pp60v-*src* in transformed cells is in the nucleus. Most of the protein is at the other extreme of the cell: it is bound to the plasma membrane, the thin film that encloses the cell and mediates its interactions with the outside world. Many cell biologists have argued that the control of cell growth may originate at the plasma membrane and its associated structures.

Inspection of the plasma membrane of cells transformed by *src* has provided the first correlation between the action of pp60v-*src* on a specific cellular protein and one of the typical changes in structure and function seen in cancer cells. By means of specialized techniques of photomicroscopy

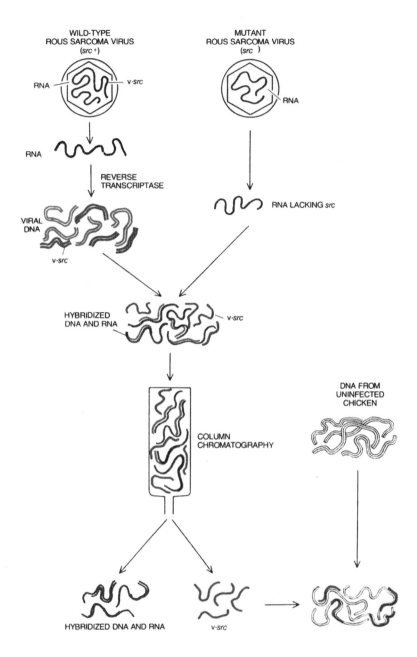

RADIOACTIVE PROBE with which to search for *src* in normal cells was prepared by Dominique Stehelin. RNA carrying the gene v-*src* was isolated from the Rous sarcoma virus (*top left*) and was copied into DNA by reverse transcriptase in the presence of radioactively labeled DNA precursors. The resulting radioactive DNA fragments were denatured (complementary strands were separated) and mixed with RNA extracted (*top right*) from a mutant virus that lacks v-*src*. Single strands of DNA will "hybridize" with closely related strands of DNA or RNA. Most of the radioactive DNA hybridized with the deletion-mutant RNA, but the *src* fragments, finding no complementary RNA, could not. The DNA-RNA hybrids were then separated by column chromatography from the unhybridized *src* DNA to yield a radioactive *src* probe. When the probe was mixed (*bottom right*) with normal chicken DNA, it hybridized, revealing the presence of a cellular proto-oncogene that was designated c-*src*.

Larry R. Rohrschneider of the Fred Hutchinson Cancer Research Center in Seattle was able to demonstrate that pp60v-*src* is concentrated in adhesion plaques: regions of the membrane that adhere to solid surfaces. In cancer cells the adhesion plaques are dismantled; the resulting decrease in cell adhesion may contribute to the ease with which most cancer cells break away from their tissue of origin and metastasize to other sites.

Rohrschneider's findings suggested that pp60v-*src* might dismantle adhesion plaques by phosphorylating one of their component proteins, or perhaps several of those proteins. Pursuing that suggestion, Sefton and S. J. Singer of the University of California at San Diego showed that pp60v-*src* phosphorylates a tyrosine unit in vinculin, a protein that is a constituent of normal adhesion plaques and becomes dispersed throughout the cell following transformation by *src*. It seems reasonable to suggest that the phosphorylation of vinculin precipitates the dismantling of adhesion plaques, but the importance of such events in the unruly behavior of cancer cells has yet to be established.

Once it was thought that the oncogenic effects of viruses might be ancillary manifestations of viral genes whose main function is to assist in the production of new virus particles. Now it is clear that the replication of retroviruses proceeds normally in the absence of oncogenes. How then can one explain the wide occurrence of oncogenes in retro-viruses and their apparent conservation in the course of evolution? A decade of investigation has furnished a surprising answer. Retrovirus oncogenes are merely cellular genes in another guise, passengers acquired from the animals in which the viruses replicate. The discovery that cells too have oncogenes has implications extending far beyond tumor virology.

## The Origin of Oncogenes

In 1972 Dominique Stehelin, Varmus and I set out to explore the "oncogene hypothesis" proposed by Robert J. Huebner and George J. Todaro of the National Cancer Institute. Seeking one mechanism to explain the induction of cancer by many different agents, Huebner and Todaro had suggested that retrovirus oncogenes are a part of the genetic baggage of all cells, perhaps acquired through viral infection early in evolution. The oncogenes would be innocuous as long as they remained quiescent. When stimulated into activity by a carcinogenic agent, however, they could convert cells to cancerous growth. We reasoned that if the hypothesis was correct, we might be able to find the *src* gene in the DNA of normal cells.

The DNA of vertebrates includes tens of thousands of genes. To search for *src* amid this vast array Stehelin fashioned a powerful tool: radioactive DNA copied solely from *src* by reverse transcriptase. The copied DNA served as a probe with which to search for cellular DNA with a nucleotide sequence similar to that of *src*. The search was carried out by molecular hybridization, in which chains of a nucleic acid (either DNA or RNA) hybridize, or form complexes, with nucleic acids to which they are related. We were exhilarated (and more than a little surprised) to learn that Stehelin's copy of *src* could hybridize with DNA from uninfected chickens and other birds. Deborah H. Spector joined us and went on to find DNA related to *src* in mammals, including human beings, and in fishes. We concluded that all

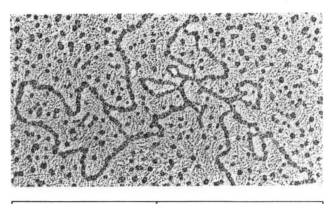

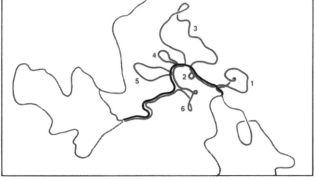

CELLULAR AND VIRAL ONCOGENES are visualized in an electron micrograph made by Richard C. Parker. Viral DNA carrying the *src* gene and chicken DNA carrying the cellular version of the gene were isolated and the double-strand DNA's were denatured. Then the single strands were allowed to hybridize with any closely related strands. In this case, as is shown in the drawing, the viral gene and the cellular one have formed a heteroduplex. (Extraneous DNA required for cloning the genes is shown in gray.) The loops in the cellular strand are six introns: intervening sequences that interrupt the protein-encoding sequences (exons) in many cellular genes but not in retrovirus genes. Such electron micrographs helped to establish that the oncogenes in cells are native to the cell and were not introduced by viruses.

vertebrates probably possess a gene related to *src,* and it therefore seemed the Huebner-Todaro oncogene hypothesis might be correct.

On closer inspection, however, the gene we had discovered in vertebrates proved not to be a retrovirus gene at all. It is a cellular gene, which is now called *c-src.* The most compelling evidence for this conclusion came from the finding that the protein-encoding information of *c-src* is divided into several separate domains, called exons, by intervening regions known as introns. A split configuration of this kind is typical of animal-cell genes but not of the genes of retro-viruses. Apart from their introns, the versions of *c-src* found in fishes, birds and mammals are all closely related to the viral gene *v-src* and to one another. It appears the vertebrate *src* gene has survived long periods of evolution without major change, implying that it is important to the well-being of the species in which it persists.

The mystery presented by *c-src* deepened with the discovery that the gene not only is present in normal cells but also is active in them, that is, the gene is transcribed into messenger RNA and the RNA is translated into protein. Molecular hybridization with Stehelin's radioactive copy of *v-src* brought the RNA to view first, in both avian and mammalian cells. The protein was more elusive, mainly because it is synthesized in very small quantities in most cells. Success came when we and others probed for the cellular protein with antibodies prepared originally for the pursuit of the viral transforming protein, pp60v-*src.* The cellular protein isolated with the aid of these antibodies proved to be virtually indistinguishable from the viral protein, and it was therefore named pp60c-*src.* The two proteins are similar in size and chemical structure; both catalyze the phosphorylation of tyrosine and both are tightly bound to the plasma membrane of cells (transformed cells in the case of pp60v-*src,* normal cells in the case of pp60c-*src*). It is as if the two proteins were designed for the same purpose, even though one is a viral protein that causes cancer and the other is a protein of normal cells.

# Cellular Oncogenes

T he findings with respect to *src* were the first hint of a generalization whose extent and significance have yet to be established. Of 17 retrovirus oncogenes identified to date, 16 are known to have close relatives in the normal genomes of vertebrate species. Most of these cellular relatives of viral oncogenes obey the principles first deduced for *c-src*. They have the structural organization of cellular genes rather than of viral genes; they seem to have survived long periods of evolution, and they are active in normal cells. To account for these facts and for the remarkable similarity between retrovirus oncogenes and their normal cellular kin most virologists have settled on the idea that retrovirus oncogenes are copies of cellular genes. It appears the oncogenes were added to preexistent retrovirus genomes at some time in the not too distant past. How and why retroviruses have copied cellular genes is not known, but there is reason to think the copying continues today, and it may even be possible to recapitulate the process in the laboratory.

The vertebrate genes from which retrovirus oncogenes apparently arose were at first called proto-oncogenes, to emphasize their evolutionary

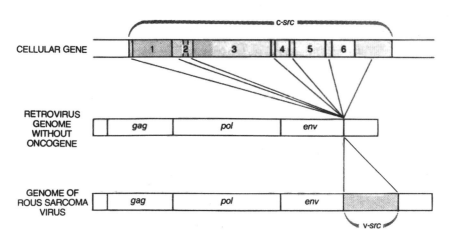

SPLIT CELLULAR GENE *c-src* (*top*) consists of exons and introns. The cellular gene was somehow picked up by a preexisting retrovirus; the introns were eliminated and the exons, spliced together, were inserted into the viral genome (*middle*) to complete the Rous sarcoma virus genome (*bottom*). In addition to *src* the genes are *gag*, which encodes the protein of the viral capsid, pol (encoding the enzyme reverse transcriptase) and *env* (encoding glycoprotein spikes of viral envelope). Other retrovirus oncogenes are thought to have similar origin.

significance and to avoid implying that the cellular genes themselves have oncogenic potential. Now it is clear that they do have such potential. They are cellular oncogenes. The investigations that justify this designation began with this question: If retrovirus oncogenes are merely copies of genes found in normal cells, how can one account for the devastating effects of the viral genes on infected cells? Two explanations have been offered. The mutational

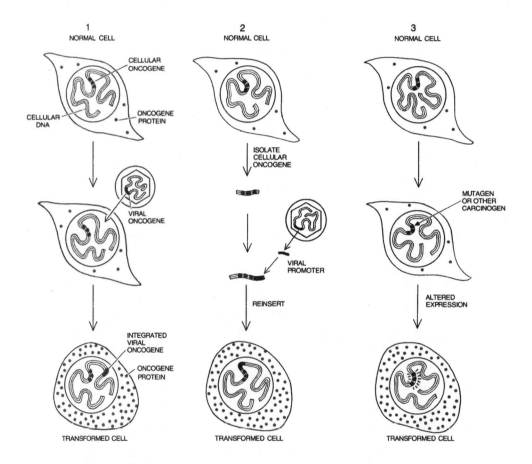

DOSAGE HYPOTHESIS holds that a cellular oncogene directs the synthesis of an amount of a normal protein product that is required for normal growth (*top row*) and that transformation to cancerous growth results from overproduction of the normal protein. In the case of infection by a tumor retrovirus (1) the overproduction is directed by the viral oncogene under viral control. The cellular oncogene itself has tumor-inducing potential, however, as has been demonstrated by a recombinant-DNA experiment (2). If a viral "promoter" is attached to the cellular oncogene and the activated form of the gene is reinserted into cells in culture, excessive quantities of the oncogene product may be synthesized, transforming the cell. A similar process may explain cancerous growth induced by a wide variety of agents other than retroviruses (3). If the cell's DNA is damaged by a mutagen or other carcinogen in some way that happens to increase the activity of the cellular oncogene, the oncogene's product may be made in excessive quantity, converting the cell to cancerous growth.

hypothesis proposes that viral oncogenes differ from their cellular progenitors in subtle but important ways as a result of mutations introduced when the cellular genes were copied into the retrovirus genome. For example, the apparently similar enzymatic activities of pp60v-*src* and pp60c-*src* might actually have different targets in the cell and might therefore have very different effects on cellular behavior. The alternative dosage hypothesis suggests that retrovirus oncogenes act by brute force, overburdening cells with too much of what are essentially normal proteins carrying out normal functions. In this view the genesis of cancer by retrovirus oncogenes is related to the amount of the viral proteins rather than to any unique properties they have.

It is too early to know which of these views is correct, but initial indications favor the dosage hypothesis. First, the doses of retrovirus transforming proteins are unquestionably large. The signals directing the activity of retrovirus genes are quite powerful, with the result that the amount of protein produced from a viral oncogene is far larger than the amount usually produced from the corresponding cellular gene; it is clearly possible that the cell might be overwhelmed. More important evidence has come from attempts to test a central prediction of the dosage hypothesis: If retrovirus oncogenes and cellular oncogenes are indeed identical in function, it should be possible to find circumstances under which the cellular genes themselves can induce cancerous growth.

## Cellular-Gene Oncogenesis

The first test of this prediction came from the remarkable experiments of Hidesaburo Hanafusa and his colleagues at Rockefeller University. Hanafusa found strains of the Rous sarcoma virus that had lost large portions of the *src* gene (but not all of it) and were therefore incapable of inducing the characteristic sarcoma in experimental animals. When Hanafusa injected the crippled viruses into chickens and then recovered the virus particles manufactured in the infected cells, he was astonished to discover that the v-*src* gene of the virus had been reconstituted. Apparently genetic material from the c-*src* gene was recombined with the viral genome while the virus was growing in the birds. The virus bearing the reconstituted gene was again fully capable of causing tumors, even though as much as three-fourths of its oncogene had just been acquired from a cellular gene. Hanafusa was able to repeat this extraordinary exercise at will, in quails as well as in chickens. His findings lent weight to the idea that the functions of c-src and v-src are the same, but many tumor virologists were unpersuaded in the absence of more direct evidence for the tumorigenic potential of the cellular genes.

Now such evidence is at hand. The research groups of George F. Vande Woude and Edward M. Scolnick of the National Cancer Institute exploited the techniques of genetic engineering to isolate three cellular oncogenes (one from mice and the other two from rats) and to show directly that these genes can induce cancerous growth in cultured cells. The feat was accomplished by attaching to the cellular genes a viral "promoter," a DNA-encoded signal that helps to regulate the expression of a nearby gene. In accordance with the dosage hypothesis, when the src-promoter complex was introduced into cells, some of the cells were transformed as if they had received a viral oncogene, whereas what they had received was a cellular gene under viral orders to work harder than usual. Moreover, cells transformed by the two rat cellular oncogenes could be shown to make very large quantities of the proteins encoded by the genes, again in accordance with the dosage hypothesis.

Why should an overabundance of a normal protein wreak such havoc? The question can be answered with assurance only when the role of cellular oncogenes in the orderly affairs of normal cells is understood. Perhaps cellular oncogenes are part of a delicately balanced network of controls that regulate the growth and development of normal cells. Excessive activity by one of these genes might tip the balance of regulation toward incessant growth.

There is evidence that the activities directed by viral and cellular oncogenes do help to control the growth of normal cells. Whereas at first the phosphorylation of tyrosine by pp60v-src seemed to be an anomalous process whose foreign nature might underlie the cancerous response to src, that view was reversed when Stanley Cohen of the Vanderbilt University School of Medicine found a role for tyrosine phosphorylation in the housekeeping of normal cells. Having discovered and purified a small "epidermal growth factor" whose binding to the surface of cells stimulates DNA synthesis and cell division, he considered how the signal for these events might be transmitted from the exterior of the cell to the interior. Cohen first showed that the binding of epidermal growth factor to cells brings about phosphorylation of proteins; prompted by the findings with pp60v-*src*, he ultimately found that the phosphorylation elicited by epidermal growth factor specifically affects tyrosine. Other workers have since shown that some proteins phosphorylated in response to epidermal growth factor can be phosphorylated by pp60v-*src*. A normal stimulant of cell division (epidermal growth factor) and an abnormal one (pp60v-*src*) thus appear to play on the same keyboard. The implication is that tyrosine phosphorylation by pp60c-*src* has a part in regulating the growth of normal cells.

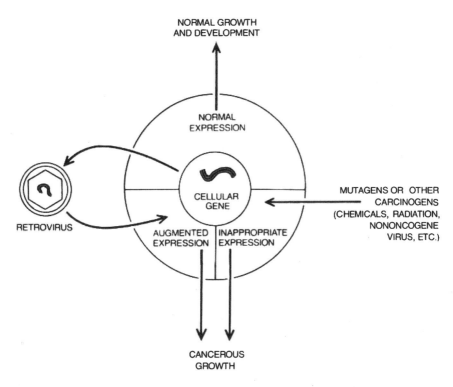

CANCER-GENE CONCEPT, supported by oncogene data and other preliminary evidence, suggests a unifying explanation for various forms of carcinogenesis. The common central element is a group of cellular genes required for normal growth and development. Transplanted into a retrovirus genome (*left*), such a gene becomes an oncogene. Cancer can also result if the cellular gene is affected by any of a wide variety of mutagens or other carcinogens (*right*).

## In Search of a Unified Theory

Retroviruses do not seem to be a major cause of human cancer. They may nonetheless have pointed the way to the central mechanisms by which the disease arises. It is generally thought that cancer begins with damage to DNA, although the exact nature of the damage is in dispute. How might the damage cause cancerous growth? Most recent efforts to answer the question in a way that might apply to all forms of cancer have invoked the existence of "cancer genes": components of the normal cellular genome whose activity is unleashed or augmented by carcinogens of various kinds and is then responsible for sustaining the undisciplined behavior of cancer cells. In this scheme cancer genes are viewed not as alien intruders but as normal, indeed essential, genes run amok; the damage done by a carcinogen turns friend into foe, perhaps by acting directly on the cancer gene or perhaps by crippling a second gene that normally polices the activity of the cancer gene.

Medical geneticists may have detected the effects of cancer genes years ago, when they first identified families whose members inherit a predisposition to some particular form of cancer. Now, it appears, tumor virologists may have come on cancer genes directly in the form of cellular oncogenes. In their viral form these genes are tumorigenic, and Vande Woude's and Scolnick's results imply that the cellular genes can also transform cells. It is therefore easy to imagine that cancer genes and the cellular oncogenes revealed by retroviruses are one and the same. The oncogene hypothesis has been restaged with the actors now cellular rather than viral genes; the dosage hypothesis serves to explain why the augmented activity of a normal cellular gene might cause cancer.

Evidence in support of these ideas comes from the study of chicken retro-viruses that induce lymphoma, a lethal tumor of the immune system. The chicken lymphoma viruses have no oncogenes. Why then do they cause tumors? William S. Hayward and Benjamin G. Neel of Rockefeller University and Susan M. Astrin of the Institute for Cancer Research in Fox Chase, Pa., have discovered that in tumors induced by the chicken lymphoma viruses the viral DNA is almost always inserted into cellular

DNA in the immediate vicinity of a single cellular oncogene (not *c-src* but a more recently recognized oncogene known as *c-myc*). As a seeming consequence of the insertion, the expression of the cellular oncogene is greatly amplified.

These findings fit the concept of cancer genes quite well. The insertion of lymphoma-virus DNA into the host genome is analogous to mutagenesis or other forms of damage introduced by carcinogens of many kinds. The insertion apparently stimulates the activity of a gene that is known to be oncogenic when it appears (as v-*myc*) in a different chicken retrovirus; the stimulated action of the cellular oncogene seems to be responsible, at least in part, for the genesis of tumors. Retroviruses without oncogenes induce a variety of tumors; by identifying the site where the viral genome is inserted into cellular DNA in some of those tumors, virologists may be able to discover cancer genes not yet identified by other means.

The unveiling of cancer genes (in the form of cellular oncogenes) by retroviruses was serendipitous. Must investigators be content with the pace at which retroviruses thus offer up new oncogenes from within the cell? Apparently not. Robert A. Weinberg of M.I.T. and Geoffrey Cooper of the Harvard Medical School have broadened the search for cancer genes beyond the purview of tumor virology. They have shown that gene-length bits of DNA isolated from some tumors (tumors that were not induced by viruses) can transmit the property of cancerous growth when they are introduced into previously normal cells in culture.

Weinberg and Cooper have evidently found a way of transferring active cancer genes from one cell to another. They have evidence that different cancer genes are active in different types of tumors, and so it seems likely that their approach should appreciably expand the repertory of cancer genes available for study. None of the cancer genes uncovered to date by Weinberg and Cooper is identical with any known oncogene. Yet it is clearly possible that there is only one large family of cellular oncogenes. If that is so, the study of retroviruses and the procedures developed by Weinberg and Cooper should eventually begin to draw common samples from that single pool.

## A Final Common Pathway

**N**ormal cells may bear the seeds of their own destruction in the form of cancer genes. The activities of these genes may represent the final common pathway by which many carcinogens act. Cancer genes may be not unwanted guests but essential constituents of the cell's genetic apparatus, betraying the cell only when their structure or control is disturbed by carcinogens. At least some of these genes may have appeared in retroviruses, where they are exposed to easy identification, manipulation and characterization.

What has been learned from oncogenes represents the first peep behind the curtain that for so long has obscured the mechanisms of cancer. In one respect the first look is unnerving, because the chemical mechanisms that seem to drive the cancer cell astray are not different in kind from mechanisms at work in the normal cell. This suggests that the design of rational therapeutic strategies may remain almost as vexing as it is today. It will be of no use to invent means for impeding the activities responsible for cancerous growth if the same activities are also required for the survival of normal cells.

However the saga of oncogenes concludes, it presents some lessons for everyone concerned with cancer research. The study of viruses far removed from human concerns has brought to light powerful tools for the study of human disease. Tumor virology has survived its failure to find abundant viral agents of human cancer. The issue now is not whether viruses cause human tumors (as perhaps they may, on occasion) but rather how much can be learned from tumor virology about the mechanisms by which human tumors arise.

| Oncogene | Species of Origin | Type of Tumor | Proto-oncogene in Vertebrate DNA | Oncogene Product | | |
|---|---|---|---|---|---|---|
| | | | | Protein Kinase | Phosphorylates Tyrosine | Located on Plasma Membrane |
| V-*src* | Chicken | Sarcoma | Yes | Yes | Yes | Yes |
| V-*fps* | Chicken | Sarcoma | Yes | Yes | Yes | Yes |
| V-*yes* | Chicken | Sarcoma | Yes | Yes | Yes | 7 |
| V-*ros* | Chicken | Sarcoma | Yes | Yes | Yes | ? |
| V-*myc* | Chicken | Carcinoma, Sarcoma, Leukemia | Yes | ? | ? | 7 |
| V-*erb* | Chicken | Leukemia, Sarcoma | Yes | 7 | ? | ? |
| V-*myb* | Chicken | Leukemia | Yes | 7 | 7 | 7 |
| V-*rel* | Turkey | Lymphoma | Yes | ? | 7 | 7 |
| V-*mos* | Mouse | Sarcoma | Yes | 7 | ? | ? |
| V-*bas* | Mouse | Sarcoma | Yes | ? | ? | ? |
| V-*abl* | Mouse | Leukemia | Yes | Yes | Yes | Yes |
| V-*ras* | Rat | Sarcoma, Leukemia | Yes | Yes | 7 | Yes |
| V-*fes* | Cat | Sarcoma | Yes | Yes | Yes | 7 |
| V-*fms* | Cat | Sarcoma | Yes | Yes | ? | 7 |
| V-*sis* | Monkey | Sarcoma | Yes | ? | 7 | 7 |

ONCOGENES OF RETROVIRUSES are listed with their species of origin and the tumors they induce, which provide experimental models for most major human cancers. Each of these 15 viral oncogenes is apparently related to a proto-oncogene found in the DNA of vertebrate animals. The product of at least eight of the genes is a protein kinase, a phosphorylating enzyme. Six of the eight kinases have so far been shown to phosphorylate the amino acid tyrosine; at least four of the kinases bind to the plasma membrane of cancerous cell.

## ABOUT THE AUTHOR

**J. Michael Bishop** (born 1936) was awarded the Nobel Prize in Physiology or Medicine in 1989 along with Harold E. Varmus for their discovery of the cellular origin of retroviral oncogenes. Bishop currently serves as an active faculty member and chancellor at the University of California, San Francisco.

# Prions

*These agents of infectious disease present
a biological conundrum: Prions contain protein
and reproduce in the living cell, yet no
DNA or RNA has been found in them.
What is the nature of their genome?*

## Stanley B. Prusiner

The nucleic acids DNA and RNA would seem to be the lowest common denominator of life. Living organisms are exceedingly diverse, both as species and as individuals, but they are all alike in having a genome of nucleic acid. In bacteria as in human beings it is DNA that specifies the structure of enzymes and other proteins and thereby determines form and development; the replication of the DNA is the crucial event in reproduction. Even viruses, which cannot reproduce independently and whose status as living organisms is therefore questionable, take their identity from a molecule of DNA or RNA. The viroids, which transmit certain plant diseases, consist of nothing but RNA. The principle that genetic information invariably flows from nucleic acids to proteins has been called the central dogma of molecular biology.

It now appears that an infectious agent named a prion may stand out as a remarkable exception to the rule that every organism carries nucleic acids defining its own identity. The prion is known to be capable of initiating the production of new prions, at least in certain mammalian cells. Moreover, among the molecular components of the prion there is at least one protein, and so one would expect to find a DNA or RNA template specifying the structure of the protein. The evidence gathered so far, however, indicates the prion has no nucleic acid at all. Even if some DNA or RNA is ultimately found in the prion, there is probably not enough to encode the structure of the protein. From these facts it does not necessarily follow that the prion violates the central dogma—the latest results favor less heretical hypotheses—but there is little question its mode of reproduction is highly unusual.

Two diseases are known to be caused by prions. They are scrapie, a neurological disorder of sheep and goats, and Creutzfeldt-Jakob disease, a rare human dementia that recently came to public notice when it was identified as the cause of the death of the choreographer George Balanchine. Prions are also considered the probable agents of two other human diseases of the central nervous system: kuru, which has been observed only among highland tribes of New Guinea, and Gerstmann-Sträussler syndrome. Prions are listed among the possible causes of several other diseases, but the evidence remains circumstantial. Perhaps the most important disease in this third category is Alzheimer's disease, the commonest form of senile dementia and the fourth leading cause of death in the U.S.

Scrapie and the other diseases in which prions are implicated are classified as "slow infections." They are characterized by a prolonged incubation period of months, years or possibly decades in which the patient or the host animal is without symptoms; once the illness begins, however, it progresses steadily and generally leads to death.

The name scrapie comes from the tendency of sheep that have the disease to scrape off much of their wool. The disorder has been known for more than 200 years, and by 1935 French workers had shown convincingly that it can be transmitted from one sheep to another by inoculation. The demonstration of transmissibility implies that there is an infectious agent capable of reproducing itself in the host animal.

Scrapie remained an obscure veterinary disorder until 1959, when William J. Hadlow of the Rocky Mountain Laboratory of the National Institute of Allergy and Infectious Diseases suggested that scrapie and kuru might be related. The first medical descriptions of kuru had been published just two years earlier by Vincent Zigas and D. Carleton Gajdusek of the National Institute of Neurological and Communicative Disorders and Stroke. They had discovered the disease in New Guinea tribes said to practice a form of ritual cannibalism in which the brain of a deceased relative is eaten as an act of homage. The practice has since ended, and the incidence of kuru has diminished.

In 1966 Gajdusek, Clarence J. Gibbs, Jr., and Michael P. Alpers reported the transmission of kuru to apes. Two years later Gajdusek and Gibbs showed that Creutzfeldt-1akob disease also can be transmitted to apes. More recently Colin L. Masters, Gajdusek and Gibbs have shown that Gerstmann-Sträussler syndrome is transmissible to apes and monkeys.

The clinical and pathological signs of scrapie, kuru, Creutzfeldt-Jakob disease and Gerstmann-Sträussler syndrome suggest they are all closely related. The initial symptoms of scrapie, kuru and Gerstmann-Sträaussler syndrome are difficulty in walking and loss of coordination, indicating a functional disorder of the cerebellum. In kuru, dementia appears late in the course of the illness. Creutzfeldt-Jakob disease generally begins as a dementia, although a few cases show early signs similar to those of kuru. In none of the diseases is there any sign of an inflammatory process or fever, and counts of cells in the cerebrospinal fluid remain normal. These last observations are evidence that the immune system does not respond to the disease agent.

Pathological changes brought on by the diseases are confined to the central nervous system. A consistent indicator is abnormal proliferation of the astrocytes, a class of supporting cells in the brain. In neurons there is a depletion of dendritic spines, which have a role in the transmission of nerve impulses. In some of the disorders numerous vacuoles give the brain tissue a spongy appearance. Amyloid plaques, which are deposits of material with a fibrillar structure, have been observed in many cases of the diseases, but they do not seem to be a constant or obligatory feature.

In the past two decades considerable effort has been given to identifying the cause of scrapie. The unusual structure and properties of the agent and the slow and laborious biological assays needed to measure its concentration have impeded the work. By 1975 more than a dozen hypotheses had been proposed for the nature of the scrapie agent; indeed, there were more hypotheses than there were experimental groups working on the disease.

In trying to isolate an infectious agent whose structure and chemistry are unknown it is necessary to take an empirical approach. Typically a tissue sample from an infected animal is homogenized and then separated into fractions that differ in some physical or chemical property. The concentration of agent in each fraction is then assayed, and the purest fraction is singled out for further attention. In the case of scrapie the only way to measure the concentration of the agent has been by detecting its ability to induce the disease in animals.

For many years all such measurements had to be done by the method of end-point titration. Animals were inoculated with progressively more dilute specimens of material; the most dilute specimen capable of inducing the disease gave a measure of the concentration of the agent in the original material. In the early work with sheep and goats an entire herd of animals and several years of observation were needed to evaluate a single sample.

In 1960 Richard Chandler succeeded in transmitting scrapie to mice. End-point titrations in mice typically employed 10 dilutions, with each dilution being weaker than the one before by a factor of 10. Six mice would be inoculated at each dilution; those receiving a large dose would become ill in from four to five months, but in the animals given the most dilute solution capable of causing disease almost a year would pass before symptoms appeared. Hence 60 mice would have to be kept for a year before the end point could be determined. Although end-point titration in mice was an improvement over work in sheep and goats, this method of measuring the scrapie agent was still slower and more cumbersome than the methods used by Pasteur in his studies of viruses almost a century earlier.

In 1978 my colleagues and I found an alternative to end-point titration. Three years earlier Richard Marsh of the University of Wisconsin at Madison and Richard H. Kimberlin had described a form of scrapie in hamsters whose onset is about twice as fast as it is in mice. Studying the hamster disease, we found strong correlations between the concentration of scrapie agent and the rapidity of disease onset and between concentration and time of death. Thus instead of determining how much a sample could be diluted and still cause disease, we measured how fast a sample with a known dilution brought on disease symptoms and caused death.

The assay based on incubation times has been found to give an accurate measure of concentration for samples with a high titer of the agent. The gains in speed and economy have had a profound effect: we estimate that they have accelerated our work by a factor of 100. Instead of observing 60 animals for a year, we can assay a sample with just four animals in 60 days.

My work on scrapie began as a collaboration at the Rocky Mountain Laboratory with Hadlow and the late Carl M. Eklund. Our initial efforts to purify and isolate the agent utilized material from the mouse spleen, which we analyzed by centrifugation and end-point titration in mice. A centrifuge separates the components of a mixture according to their size and density. In the method we adopted a sample was spun at a particular speed for a specified interval, then the bioassay was carried out to determine how much of the infectious agent had sedimented and how much had remained in the supernatant fluid. The procedure was done for a wide range of speeds and times in a series of experiments that took almost two years. When we had finished, we repeated the study to make certain our results were; reproducible.

In these early studies the greatest purification attained was about a thirtyfold enrichment of scrapie agent. One of the factors limiting the degree of purification was also one of our major findings: the infectious particles were shown to be extremely heterogeneous in size and density. Judging from the rate of sedimentation in the centrifuge, some of the particles were almost as large as mitochondria or bacteria; others seemed to be substantially smaller than the smallest viruses. The broad range of sizes implied that the scrapie agent can exist in many molecular forms; the observations could be explained, for example, if very small infectious particles aggregate to form much larger clusters. In an attempt to make more homogeneous preparations, as well as to facilitate the separation of the agent from cellular molecules, we added detergents. Although the detergents we employed had little effect on the observed heterogeneity, they did aid in purification.

The effort to purify the scrapie agent continued in my laboratory at the School of Medicine of the University of California at San Francisco. The story of our endeavor can be told briefly, but it represents a decade of painstaking and sometimes frustrating labor. A large number of people have made important contributions. Among them I should like to mention Richard Baringer, Ronald Barry, Paul Bendheim, David Bolton, Karen Bowman, Patricia Cochran, Steve DeArmond, Darlene Groth, Michael McKinley and Daniel Stites.

It was at this point that we switched from mice to hamsters; in addition to providing a faster assay, the hamster brain has a titer of scrapie agent 100 times greater than that of the mouse spleen. Our purification method again began with detergent extraction and centrifugation, but we added three more steps: exposure to nucleases, exposure to proteases and analysis by gel electrophoresis. We had found that the infectivity of the scrapie agent is unaltered by digestion with nucleases, which are enzymes that catalyze the breakdown of nucleic acids. Hence treating the preparation with a nuclease eliminated most cellular nucleic acids while leaving the scrapie agent intact. Proteases, which cut the chain of amino acids in a protein, were used in a similar way to remove extraneous proteins. The final step in our revised procedure, gel electrophoresis, separates molecules according to the rate at which they migrate through a gel under the influence of an applied electric field. The rate at which a molecule moves is determined primarily by its electric charge, although size and shape also have an effect.

The purification method culminating in electrophoresis gave an overall purification by a factor of about 100, which was enough to establish several vital facts. The most important result was a convincing demonstration that the biological activity of the scrapie agent depends on a protein. We therefore introduced the term prion, for "proteinaceous infectious particle."

We have since returned to centrifugation as the primary method of purification, but the technique is somewhat different from the one employed in the early studies. The specimen is placed on top of a sucrose solution arranged to form a gradient of progressively higher density. When the centrifuge is spun, components of the specimen migrate through the gradient until they reach a level where their own density matches that of the surrounding solution.

Our first version of the sucrose-gradient separation gave a thousand-fold purification of the scrapie agent. We were thereby able to demonstrate that the bulk of the protein consists of a single molecular species, which we designated PrP, for prion protein. A larger-scale version of the purifi-

| Disease | Caused by Prions? | Naturalhost Species | Experimentalhost Species | Incubation Period |
|---------|-------------------|---------------------|--------------------------|-------------------|
| Scrapie | Yes | Sheep, Goats | Mice, Hamsters, Monkeys | 2 Months to 2 Years or More |
| Creutzfeldt-Jacob Disease | Yes | Human Beings | Apes, Monkeys, Mice, Goats, Guinea Pigs | 4 Months to 20 Years or More |
| Kuru | Probably | Human Beings | Apes, Monkeys | 18 Months to 20 Years or More |
| Gerstmann-Straussler Syndrome | Probably | Human Beings | Apes, Monkeys | 18 Months or More |
| Transmissible Mink Encephalopathy | Probably | Mink | Monkeys, Goats, Hamsters | 5 Months to 7 Years or More |
| Chronic Wasting Disease | Probably | Mule Deer, Elk | Ferrets | 18 Months or More |

PRION DISEASES are classified in two categories according to the strength of the evidence implicating prions in their causation. Scrapie is a prion disease by definition, and there is substantial experimental evidence that prions also bring on Creutzfeldt-Jakob disease. Prions are considered the probable cause of at least four more diseases, including two human neurological disorders. The four diseases are known to be transmissible, their symptoms resemble those of scrapie and Creutzfeldt-Jakob disease, and they all induce similar changes in brain tissue.

cation procedure was then developed, based on a centrifuge with a "zonal rotor," a large vessel in which substantial quantities of material could be separated in a sucrose gradient. This latest separation technique yields an enrichment of 5,000 times.

With the greater purification it was possible to show that the most infectious fractions of the centrifuged material include essentially one protein, namely PrP. Further studies of the protein by electrophoresis have shown that it has an apparent molecular weight of between 27,000 and 30,000. We now have reason to think this estimate is probably too high. We have found that PrP is a glycoprotein, that is, a protein in which sugars are bound to the amino acid units. Using electrophoresis to measure the molecular weight of a glycoprotein often gives an erroneously high result. In any event, PrP is a comparatively small protein, less than half the size of hemoglobin.

Electron micrographs of the purified material have revealed numerous rod-shaped particles much too large to be the individual prions. The rods had been observed in earlier experiments, but we had been unable to verify the obvious hypothesis that they are aggregates of prions; the possibility remained that they were not the disease agent itself but the product of some pathological change brought on by the disease. With our highly purified fractions we were able to establish that the rods are composed of PrP and hence can be considered prion aggregates.

The isolation of reasonably pure specimens in substantial quantity has proved to be the key to answering many puzzling questions about the prion. Quite recently it has enabled us to raise antibodies to prions in experimental animals, a feat we had been attempting for years without success. The crucial factor turned out to be the quantity of material injected into the animal, which we were able to increase by a factor of 10 (to roughly 100 micrograms). The availability of antibodies can be expected to work a great change in the pace and the technology of prion research. For example, it may greatly reduce the need for biological assays. The concentration of prions in a sample could be determined in hours rather than months by measuring the affinity of the antibodies for components of the material. The same affinity could be the basis of a new purification technique.

Study of the purer specimens has raised new questions as well as answering old ones. We have noted an extraordinary resemblance between dense collections of prion rods and the amyloid plaques seen in some cases of scrapie and Creutzfeldt-Jakob disease. It would seem reasonable to suppose the amyloid plaques are in fact deposits of prions in an aggregated state. The trouble is that amyloid plaques are also characteristic of disorders in which there is no strong reason to suspect a prion infection. The plaques are closely associated with Alzheimer's disease, which has not been shown to be transmissible.

Our conclusion that a protein is a component of the prion and is necessary for infectivity emerged directly from our work on purification. In the course of that work we examined numerous chemical reagents to determine whether or not they alter the infectivity of the prion. The essence of our findings is that substances known to disrupt proteins diminish prion infectivity. Substances that have no effect on proteins leave the infectivity unchanged.

The clearest evidence comes from experiments with proteases. A protease is highly specific: it has almost no effect on biological molecules other than proteins. In early studies with impure material, protease treatment gave equivocal results, but when enriched fractions became available, we were able to show convincingly that proteases reduce prion infectivity. Furthermore, the confusion over the early results has been resolved. It turns out that PrP is resistant to protease treatment compared with most cellular proteins. Consequently in an impure mixture with many other proteins most of the enzyme activity is directed toward competing substrates. We have exploited this effect in the purification of prions. Early in the procedure extraneous proteins can be digested with a protease without significantly degrading the prion protein.

Several other reagents damage proteins not by cutting the chain of amino acids but by unfolding it; the protein is said to be denatured. The detergent sodium dodecyl sulfate (SDS) is such a substance, and boiling a prion solution in SDS abolishes its infectivity. Phenol, urea and certain salts are also denaturing reagents; they too reduce the biological activity of prions.

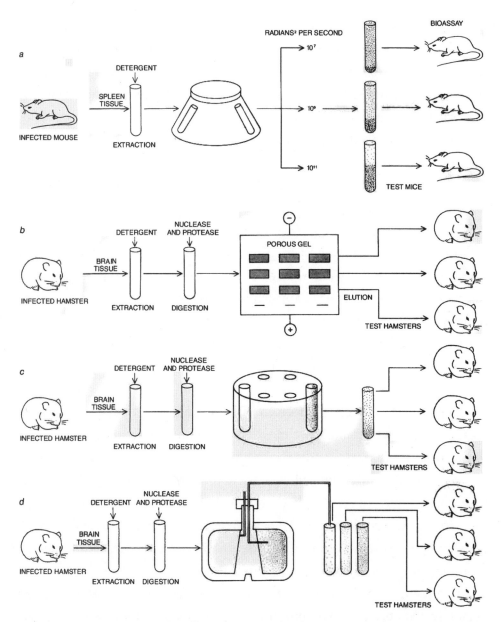

PURIFICATION TECHNOLOGIES employed by the author and his colleagues in the past 10 years have yielded progressively more concentrated solutions of prions. Detergent extraction and centrifugation with a fixed-angle rotor (*a*) separated a specimen into fractions according to particle size and achieved a purification of 30 times. Gel electrophoresis (*b*) separates the components of a solution according to their electric charge and size. Electrophoresis in conjunction with detergent extraction and enzyme digestion yielded a hundredfold purification and led to the demonstration that a protein is essential for prion infectivity. In sucrose-gradient centrifugation molecules migrate through layers of sugar solution until they reach a level matching their own density. Small-scale experiments were done with a vertical-tube rotor (*c*). The resulting thousandfold purification led to the identification of PrP. Purifications by a factor of 5,000 are now being achieved by sucrose-gradient centrifugation in a zonal rotor (*d*), which can process much larger quantities. With more material of higher purity it has been possible to investigate the composition and structure of PrP and also to raise antibodies that specifically bind to it.

All the reactions described above are irreversible. Another reagent, diethyl pyrocarbonate (DEP), modifies proteins chemically in a way that can be reversed by exposure to hydroxylamine. As in the case of proteases, DEP gave inconsistent results in early experiments with impure mixtures; we have since shown that it does diminish the titer of infectious agent in purified fractions. The infectivity can be restored by treatment with hydroxylamine.

The presumed target of all these treatments is PrP. The possibility remained, however, that PrP was not a structural component of the prion but a pathological product of the scrapie infection. We therefore undertook a lengthy series of experiments, which we feel has established that the prion is indeed composed at least in part of PrP molecules.

The first line of evidence is simply that PrP has been found in every sample with a high titer of prions, whether it was prepared by centrifugation or by electrophoresis and even if it was purified from scrapie-infected tissue before the appearance of pathological changes. The concentration of PrP is directly proportional to the titer of prions. The rate of digestion by proteases is the same for PrP and for prions, and other experimental procedures that alter the prion titer also change the concentration of PrP. DEP, which we had already shown to be capable of inactivating prions, was found to bind directly to PrP. Finally, in purified material treated with protease, PrP is the only protein that can be detected, suggesting prions have just one major protein.

Having established the role of a protein in the infective process, we began to search for nucleic acid. The effort encountered consistent frustration. Almost 20 years ago the British investigators Tikvah Alper, David Haig and Michael Clarke irradiated crude homogenates of scrapie-infected tissue with both ultraviolet radiation and shorter-wavelength ionizing radiation. In general, ionizing radiation destroys cells and viruses through damage to nucleic acids; the probability of damage is roughly proportional to the size of the target molecules. Alper and her colleagues found that extremely high doses of radiation were needed to inactivate the scrapie agent. They concluded that the agent has no nucleic acid and is considerably smaller than a virus. Their conclusions were met with great skepticism, but in collaboration with James Cleaver of the School of Medicine of the University of

California at San Francisco we have since repeated the· ultraviolet-irradiation studies using purified preparations and have obtained similar results.

In many experiments we have attempted to inactivate prions by means of treatments that chemically attack nucleic acids. Experiments directly analogous to those with proteases employ a nuclease. We have exposed prion

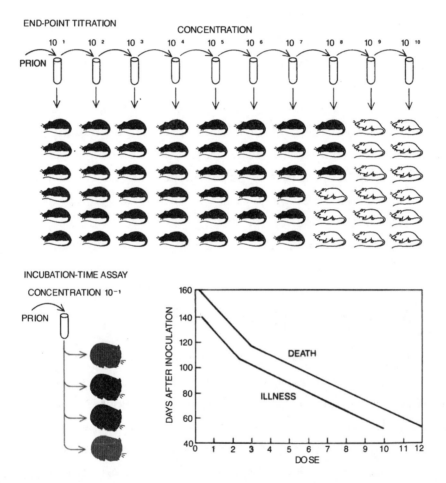

BIOLOGICAL ASSAYS OF INFECTIVITY have long been the only available methods of measuring the concentration of scrapie prions in a specimen. In the method of end-point titration 10 samples were prepared, with each sample being more dilute than the one before by a factor of 10. Each sample was injected into six mice, which were observed for up to a year for neurological signs of scrapie. The greatest dilution capable of causing disease gave a measure of the concentration of scrapie agent in the original specimen. The incubation-time method has reduced both the time and the number of animals needed to assay a specimen. It is based on the observation that the interval between inoculation and the first appearance of neurological signs as well as the interval between inoculation and death depends on the dose of agent injected. For highly purified samples measuring the time between inoculation and the onset of illness in four hamsters can indicate the concentration of scrapie agent in as little as 60 days.

solutions to several nucleases, including enzymes that destroy both DNA and RNA, without detecting any significant diminution of prion infectivity.

One possible objection to such experiments is that the enzyme may not be able to gain access to the nucleic acids; many viruses are resistant to nucleases because the protein coat protects the DNA or RNA. In collaboration with John E. Hearst of the University of California at Berkeley we tried treating prions with molecules called psoralens, which can pass through the protein coat of most viruses; on exposure to ultraviolet radiation the psoralens bind to the nucleic acid and inactivate it. Again we observed no loss of prion infectivity. Zinc ions, which catalyze the break-down of RNA, were likewise ineffective.

Our work with DEP adds further evidence. DEP can inactivate nucleic acids as well as proteins, but only proteins can be restored to functionality by hydroxylamine. Hence the recovery of infectivity we observed after hydroxylamine treatment argues that DEP is not acting on a nucleic acid.

In collaboration with Theodor O. Diener of the U.S. Department of Agriculture we have compared the effects various reagents have on prions and on viroids. Because both entities seem to be much smaller than viruses, it was once thought prions might be similar to viroids in structure, that is, they might consist of "naked" RNA. Actually the two kinds of infectious agent are antithetical. Procedures that modify proteins inactivate prions but have no effect on viroids; treatments that attack nucleic acids destroy viroids but not prions.

The question of the prion's size also bears on the nucleic acid issue. Individual prions seem to be very small. Therefore the amount of nucleic acid a prion could encapsulate is probably quite limited. The target-size studies done by Alper and her colleagues suggested the infectious scrapie particle might have a molecular weight of between 60,000 and 150,000. The remarkable heterogeneity of prions has made it difficult to determine the size of the smallest infectious particle by more direct methods. After attempting to break up aggregates of prions with detergents and other chemicals, we have investigated the size of the particles by sucrose-gradient centrifugation, by timing their passage through a chromatographic column

| TREATMENT | EFFECT ON NUCLEIC ACIDS | EFFECT ON PROTEINS | EFFECT ON PRIONS |
|---|---|---|---|
| PROTEASE | NONE | DIGESTED | LOSS OF INFECTIVITY |
| SODIUM DODECYL SULFATE (SDS) | NONE | DENATURED | LOSS OF INFECTIVITY |
| PHENOL | NONE | DENATURED | LOSS OF INFECTIVITY |
| NUCLEASE | DIGESTED | NONE | REMAIN INFECTIVE |
| ULTRAVIOLET RADIATION | DAMAGED | NONE | REMAIN INFECTIVE |
| ZINC IONS | DIGESTED | NONE | REMAIN INFECTIVE |
| PSORALEN PHOTOADDUCTS | CHEMICALLY MODIFIED | NONE | REMAIN INFECTIVE |
| HYDROXYLAMINE | CHEMICALLY MODIFIED | NONE | REMAIN INFECTIVE |
| DIETHYL PYROCARBONATE (DEP) | CHEMICALLY MODIFIED | CHEMICALLY MODIFIED | LOSS OF INFECTIVITY |
| HYDROXYLAMINE AFTER DEP | INACTIVATED | MODIFICATION REVERSED | INFECTIVITY RESTORED |

CHEMICAL NATURE OF THE PRION has been investigated by means of treatments that inactivate either proteins or nucleic acids. Proteases (enzymes that cut the chain of amino acids in a protein) can drastically reduce the infectivity of a purified solution of prions; so can certain detergents and other reagents that denature a protein by unfolding it. In contrast, nucleases (which digest the nucleic acids DNA and RNA) have no effect on prion infectivity. Prions are also highly resistant to ultraviolet radiation and zinc ions, which act mainly on nucleic acids. A nucleic acid in the prion might be protected by a coat of protein, but psoralens and hydroxylamine, which can penetrate the protein coat of many viruses, are also ineffective. The reagent diethyl pyrocarbonate (DEP) can modify both proteins and nucleic acids, but its action on proteins can be reversed by treatment with hydroxylamine, whereas its effect on nucleic acids is irreversible. Prions are inactivated by DEP and their infectivity is restored by hydroxylamine. All the chemical and physical evidence suggests the prion consists of a protein without nucleic acid, although the possibility of a small DNA or RNA cannot yet be excluded.

and by passing them through a membrane filter with pores of a known size. All the studies have given results consistent with a molecular weight of between 50,000 and 100,000, but each method has potential pitfalls. Because of the many sources of uncertainty, the most that can be said for now is that the smallest infectious form of the prion may be 100 times smaller than the smallest viruses.

If the prion has a molecular weight of 50,000, its diameter would be about five nanometers, or five billionths of a meter. If it is constructed like a conventional virus, it might take the form of an approximately spherical shell of protein surrounding a core of nucleic acid. The shell could not be less than about a nanometer thick, which would leave room in the core for no more than about 12 nucleotides. Limits on the size of any prion nucleic acid can also be derived from other measurements. The prion's resistance to inactivation by ultraviolet radiation is consistent with a nucleic acid made up of from 12 to 50 nucleotides; our experiments with psoralens would not have detected a nucleic acid with fewer than 40 nucleotides.

The failure to detect a nucleic acid in prions cannot be taken as proof that it does not exist. It could still be concealed in some way by a surrounding structure or could be present in quantities too small to be detected. Nevertheless, it seems reasonable to suggest that if the prion has any nucleic acid at all, it is likely to be less than 50 nucleotides long. In the standard genetic code three nucleotides are needed to specify each amino acid, and so the putative prion "genome" could not encode a protein with more than about a dozen amino acids. It should be noted that the molecular weight of PrP implies it has roughly 250 amino acids.

There is only one way to establish with certainty that the prion consists of nothing but protein: determine the complete sequence of amino acids in PrP, synthesize an artificial protein with the same sequence and then demonstrate that it has the same biological action as the natural protein. Surprisingly, in recent months the prospect of such a demonstration has begun to seem less remote, again because of the availability of purified preparations in large quantities. In collaboration with Leroy E. Hood of the California Institute of Technology and Stephen B. H. Kent of Molecular Genetics, Inc., we have identified the first 15 amino acids of PrP.

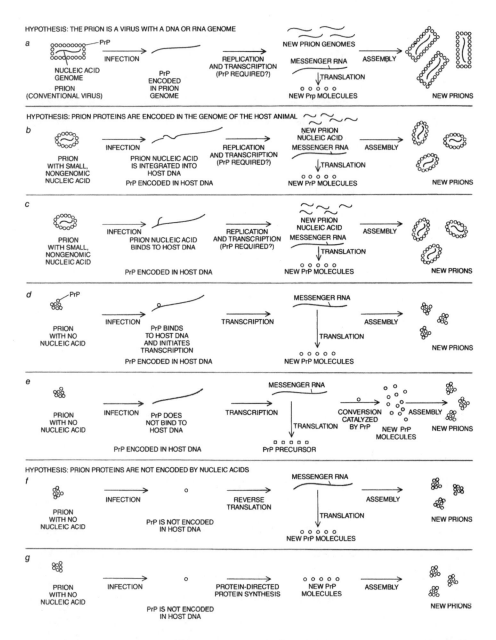

PRION REPLICATION might be accomplished by any one of several hypothetical mechanisms. The possibility that the prion is a conventional virus with a DNA or RNA genome encoding the structure of PrP (*a*) is unlikely. In another hypothesis PrP is encoded by genes in the host animal. A small prion nucleic acid might be incorporated into the host genome (*b*) or might simply bind to the host DNA (*c*); in either case it could promote the expression of PrP genes. Alternatively, if the prion has no nucleic acid at all, PrP itself might stimulate gene expression by binding to the host DNA (*d*). The requirement of biologically active PrP molecules for the synthesis of new prion particles could also be explained if PrP catalyzes the conversion of a precursor molecule into PrP (*e*). Two further possibilities violate the "central dogma" of molecular biology, which holds that genetic information flows from nucleic acids to proteins. Reverse translation might create a nucleic acid sequence based on the sequence of amino acids in PrP (*f*), or PrP might be a template for its own synthesis (*g*).

The partial sequence of PrP was determined by repeatedly employing a series of reagents that cut off the terminal amino acid of a protein, so that the amino acids in the chain are liberated one by one. In about half of the cleavage cycles we detected multiple amino acids, with one strong signal being accompanied by weaker ones. Initially we thought the minor signals indicated the existence of a spectrum of PrP molecules with slightly different amino acid sequences. We subsequently found, however, that the variant sequences differ from the major sequence only in their starting point. One minor sequence can be brought into alignment with the major sequence by moving the minor sequence forward four amino acid positions; in another case the minor sequence must be moved backward two units. The variations are observed because the PrP molecules have "ragged ends," presumably caused by treatment with proteases during purification.

Knowing even a small part of the amino acid sequence of PrP has opened up two important avenues of research. First, synthetic chains of amino acids that match the known PrP sequence have been constructed for use as antigens. Antibodies formed against the synthetic molecules can ultimately be employed in the purification of prions and in assays of their concentration as well as in probing their structure. Second, synthetic segments of nucleic acid that encode the known part of the amino acid sequence

### AMINO-TERMINAL AMINO ACIDS

| SEQUENCING CYCLE | 1 | 2 | 3 | 4 | 5 | 6 | 7 | 8 | 9 | 10 | 11 | 12 | 13 | 14 | 15 |
|---|---|---|---|---|---|---|---|---|---|---|---|---|---|---|---|
| MAJOR SIGNAL | | Gly | Gln | Gly | Gly | Gly | Thr | His | Asn | Gln | Trp | Asn | Lys | Pro | Ser | Lys |
| MINOR SIGNALS | | | | | Thr | His | Asn | | Trp | | Lys | Pro | | | | |
| | | | | Pro | Trp | | Gln | | | | Thr | His | | Gln | Trp | |

### INTERPRETED AMINO ACID SEQUENCE

| | | | | | | | | | | | | | | | |
|---|---|---|---|---|---|---|---|---|---|---|---|---|---|---|---|
| | | Gly | Gln | Gly | Gly | Gly | Thr | His | Asn | Gln | Trp | Asn | Lys | Pro | Ser | Lys |
| | | | * | * | * | Thr | His | Asn | * | Trp | * | Lys | Pro | | | |
| * | * | Pro | Trp | * | Gln | * | * | * | Thr | His | * | Gln | Trp | | | |

AMINO ACID SEQUENCE of PrP has come under investigation with the availability of substantial quantities of highly purified material. The sequence of the first 15 amino acids at the end of the molecule known as the amino terminus has been determined. Variant sequences can be brought into alignment with the sequence deduced from the major signal and interpreted as "ragged ends" caused by protease treatment in the course of the material's preparation.

of PrP have been constructed. They are being used as probes for natural DNA that encodes PrP.

Surely the preeminent question about the prion is how it replicates. There are essentially three categories of possible replication mechanisms.

The first hypothesis is that the prion, in spite of all indications to the contrary, is a conventional virus with a genome of DNA or RNA that encodes the entire structure of the prion protein. On entering a cell the nucleic acid genome of the prion would be transcribed into messenger-RNA molecules that would serve as templates for the synthesis of prion proteins. The fact that a prion protein is obligatory for scrapie infection could be explained by assuming the prion is a "negative-strand virus," one in which the viral nucleic acid alone is not enough to cause disease; a viral protein, acting as an enzyme, is needed for transcription of the viral nucleic acid into messenger RNA. The virus hypothesis is quite unlikely in view of all the information we now have about the chemical and physical structure of prions.

In one sense the conventional-virus explanation is the most conservative hypothesis; the most radical idea is that the amino acids in PrP somehow specify their own sequence during prion replication. This might come about indirectly through the "reverse translation" of the protein into RNA or DNA, which would then be interpreted by the cellular apparatus in the usual way to make more protein. Such a process has never been observed, and it would be a clear violation of the central dogma, which again states that the flow of information in the cell is always from nucleotides to proteins.

One can also imagine a mechanism in which the amino acid sequence of PrP would serve directly as a template for the construction of a new protein molecule. Such protein-directed protein synthesis has never been seen, and enzymes capable of building a complex protein by this means are not known.

The third category of possible replication mechanisms includes the ones I consider the most plausible. In these schemes there does exist a DNA gene encoding the amino acid sequence of the prion protein. The gene is not carried by the prion, however; instead it is a component of the normal mammalian genome. Infection by prions would somehow activate the gene or perhaps alter it.

If the prion includes a small piece of nucleic acid, that could be the trigger for gene activation. This hypothetical small prion nucleic acid might be inserted into a host-cell chromosome just "upstream" of the PrP gene, or in other words just ahead of the point at which transcription of the gene begins. The inserted sequence could then serve as a promoter or enhancer of gene expression. Alternatively, if the prion consists exclusively of protein, PrP itself might bind to cellular DNA in a region that controls the transcription of the PrP gene. Most proteins that bind to DNA tend to repress gene expression, but the phenomenon of a protein that stimulates its own synthesis is not entirely without precedent.

One objection to the proposal that PrP is encoded by a host gene centers on the observation that there seem to be various "strains" of prions. If replication of the disease agent is nothing more than activation of a host gene, how can the same genetic line of animals serve as host to multiple prions? Any answers suggested now are necessarily speculative, but one possibility involves the rearrangement of genes. In the synthesis of immunoglobulins, for example, the reshuffling of genes gives rise to an enormous diversity of proteins.

The question of whether or not there is a host-animal gene for PrP is likely to be settled in the coming months. From the partial amino acid sequence of PrP we have been able to prepare DNA with a sequence of nucleotides complementary to one that would encode the known part of the protein. The synthesized DNA should bind to any DNA with the complementary sequence, and so it can be employed as a probe to find a PrP gene in the cell. If prion proteins are indeed encoded by host genes, it may be more appropriate to speak of the synthesis of new prions as amplification rather than replication.

**B**ecause electron microscopy has revealed much about the structure and assembly of virus particles, many investigators have employed the electron microscope to search for a specific particle associated with scrapie infection. Spherical and cylindrical particles have been described in tissue sections and extracts. H. K. Narang found rod-shaped particles in sections of scrapie-infected brain tissue and showed that the particles are stained by substances that selectively bind to sugars. The latter findings are of notable interest, since PrP aggregates to form rods and is now known to be a glycoprotein. Henryk M. Wisniewski and his colleagues at the Downstate Medical Center of the State University of New York have found long fibrils in brain tissue infected with scrapie and Creutzfeldt-1akob disease. They believe that the fibrils can be distinguished from amyloid, that they represent a filamentous animal virus causing scrapie and that they are an elongated form of prion rods.

The prion rods have recently given us further clues to the biological and medical significance of prions. The rods can be found in preparations consisting of one protein, namely PrP, and so they must be composed largely of PrP molecules. Our recent work with antibodies to PrP confirms that identification: the antibodies specifically bind to the rods. In electron micrographs the rods are typically from 10 to 20 nanometers in diameter and from 100 to 200 nanometers long, which suggests a single rod may consist of as many as 1,000 PrP molecules, probably stacked in a crystalline array.

Perhaps the most important and intriguing aspect of the prion rods is their resemblance to amyloid. The standard procedure for identifying amyloid is to stain a section of tissue with the dye Congo red. The amyloid binds the dye and appears red in light micrographs; in addition, when the stained amyloid is viewed through polarizing filters, it exhibits the optical property called birefringence, changing from green to gold as the orientation

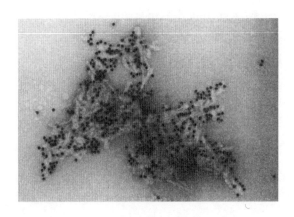

CLUSTER OF PRION RODS is seen at a magnification of 100,000 times in an electron micrograph. Adhering to the rods are antibody molecules with a specific affinity for PrP. The antibodies have been labeled with colloidal gold beads, which appear in the image as conspicuous black dots. The structure of such a cluster of rods is much like that of an amyloid plaque.

of the filters is changed. With George G. Glenner of the National Institute of Arthritis, Metabolism, and Digestive Diseases we stained clusters of prion rods with Congo red and examined them microscopically. They were red under ordinary illumination and showed green-gold birefringence between polarizing filters.

For more than 60 years amyloid plaques in the central nervous system were considered accumulations of waste material formed as a result of some disease process. Our findings suggest a quite different interpretation, namely that the plaques may be aggregations of prions in an almost crystalline state. They could be analogous to the inclusion bodies characteristic of many viral infections; inclusion bodies are crystalline arrays of virus particles.

The recent production of antibodies to PrP has allowed us to demonstrate that amyloid plaques in the brain of scrapie-infected hamsters contain prion proteins. Brain sections from the hamsters were stained first with the antibodies and then with Congo red. The same structures that bind the PrP antibodies also stain with the dye and exhibit green-gold birefringence. In collaboration with David T. Kingsbury of the University of California at Berkeley and the U.S. Naval Biosciences Laboratory we have also employed the antibodies to scrapie PrP to identify similar prion proteins in Creutzfeldt-Jakob disease. Proteins with the properties of PrP were found in the brain of both animals and human patients infected with the agent of Creutzfeldt-Jakob disease. These proteins too aggregate to form rod-shaped particles that have the characteristics of amyloid.

The disorder in which amyloid is an almost definitive sign is Alzheimer's disease; indeed, as the number of amyloid plaques increases, the degree of mental dysfunction rises. The possibility that Alzheimer's disease might be caused by prion infection is intriguing, but there is no reliable evidence that the disorder is transmissible or communicable. It is not spread by person-to-person contact. Gajdusek and Gibbs have repeatedly attempted to transmit the disease to experimental animals by inoculation; their results have been negative except in two possible cases, which they have not been able to reproduce.

If Alzheimer's disease is in fact caused by prions, two hypotheses might be proposed to explain the apparent failure of the transmissibility experiments. First, the infectious agent may not replicate in the species chosen for the experiments. Second, the incubation period may simply be too long for the disease to have been detected in the experiments; indeed, it may be longer than the lifetime of the animals. The second hypothesis is consistent with reports of an incubation period of two or three decades for kuru and Creutzfeldt-Jakob disease; it is also notable that Alzheimer's disease is commonest in older people and that the incidence rises with age. Of course, the possibility also remains that Alzheimer's disease is not caused by prions or any other infectious agent; many other causative mechanisms have been proposed.

The transmissibility of Creutzfeldt-Jakob disease raises questions of a different kind. That the disease can be transmitted in laboratory experiments is not in doubt, but it is not clear how the infection persists in a natural population. Creutzfeldt-Jakob disease has an incidence of about one case per million population throughout the world, which would not seem to be enough to sustain a chain of transmission from person to person.

Perhaps these puzzles will be resolved when the biochemistry of the prion is understood in more detail. If the prion is indeed a single protein and the product of a gene native to the host organism, the time may have come for a reconsideration of what is meant by the concept of infection.

## ABOUT THE AUTHOR

**Stanley B. Prusiner** (born 1942) was awarded the Nobel Prize in Physiology or Medicine in 1997 for his discovery of Prions - a new biological principle of infection. He is currently the director of the Institute for Neurodegenerative Diseases at University of California, San Francisco.

# How LDL Receptors Influence Cholesterol and Atherosclerosis

*The receptors bind particles carrying cholesterol and remove them from the circulation. Many Americans have too few LDL receptors, and so they are at high risk for atherosclerosis and heart attacks.*

## Michael S. Brown and Joseph L. Goldstein

Half of all deaths in the U.S. are caused by atherosclerosis, the disease in which cholesterol, accumulating in the wall of arteries, forms bulky plaques that inhibit the flow of blood until a clot eventually forms, obstructing an artery and causing a heart attack or a stroke. The cholesterol of atherosclerotic plaques is derived from particles called low-density lipoprotein (LDL) that circulate in the bloodstream. The more LDL there is in the blood, the more rapidly atherosclerosis develops.

Epidemiologic data reveal the surprising fact that more than half of the people in Western industrialized societies, including the U.S., have a level of circulating LDL that puts them at high risk for developing atherosclerosis. Because such concentrations are so prevalent, they are considered "normal," but clearly they are not truly normal. They predispose to accelerated atherosclerosis and heart attacks or strokes.

What determines the blood level of LDL, and why is the level dangerously high in so many Americans? Some answers are emerging from studies of specialized proteins, called LDL receptors, that project from the surface of animal cells. The receptors bind LDL particles and extract them from the fluid that bathes the cells. The LDL is taken into the cells and broken down, yielding its cholesterol to serve each cell's needs. In supplying cells with cholesterol the receptors perform a second physiological function, which is critical to the development of atherosclerosis: they remove LDL from the bloodstream.

The number of receptors displayed on the surface of cells varies with the cells' demand for cholesterol. When the need is low, excess cholesterol accumulates; cells make fewer receptors and take up LDL at a reduced rate. This protects cells against excess cholesterol, but at a high price: the reduction in the number of receptors decreases the rate at which LDL is removed from the circulation, the blood level of LDL rises and atherosclerosis is accelerated.

We have proposed that the high level of LDL in many Americans is attributable to a combination of factors that diminish the production of LDL receptors. Recognition of the central role of the receptors has led to a treatment for a severe genetic form of atherosclerosis, and it has also shed some light on the continuing controversy over the role of diet in atherosclerosis in the general population.

The story begins with the discovery of LDL receptors in 1973 in our laboratory at the University of Texas Health Science Center at Dallas. We were studying tissue cultures of the human skin cells called fibroblasts. Like all animal cells, cultured fibroblasts need cholesterol as a major building block of their surface membrane (the plasma membrane); they had been shown to get the cholesterol by extracting it from lipoproteins in the serum of the culture medium. There is a mixture of various lipoproteins in human serum, but we found that the fibroblasts derive most of their cholesterol from a particular lipoprotein: LDL. We were able to attribute this to the presence on the cells of highly specific receptor molecules that bind LDL and related lipoproteins.

LDL is a large spherical particle whose oily core is composed of some 1,500 molecules of the fatty alcohol cholesterol, each attached by an ester linkage to a long-chain fatty acid. This core of cholesteryl esters is enclosed in a layer of phospholipid and unesterified cholesterol molecules. The phospholipids are arrayed so that their hydrophilic heads are on the outside, allowing the LDL to be dissolved in the blood or intercellular fluid. Embedded in this hydrophilic coat is one large protein molecule designated apoprotein B-100.

It is apoprotein B-I00 that is recognized and bound by the LDL receptor, a glycoprotein (a protein to which sugar chains are attached). The receptor spans the thickness of the cell's plasma membrane and carries a binding site that protrudes from the cell surface. Binding takes place when LDL is present at a concentration of less than $10^{-9}$ molar, which is to say that the receptor can pick out a single LDL particle from more than a billion molecules of water. The receptor binds only lipoproteins carrying apoprotein B-100 or a related protein designated apoprotein E.

How is LDL taken into the cell? Our collaborator Richard G. W. Anderson discovered in 1976 that the receptors are clustered in specialized regions where the cell membrane is indented to form craters known as coated pits (because the inner surface of the membrane under them is coated with the protein clathrin). Within minutes of their formation the pits pouch inward into the cell and pinch off from the surface to form membrane-bounded sacs called coated vesicles; any LDL bound to a receptor is carried into the cell. (Receptor-mediated endocytosis, the term we and Anderson applied to this process of uptake through coated pits and vesicles, is now recognized as being a general mechanism whereby cells take up many large molecules, each having its own highly specific receptor.)

Eventually the LDL is separated from the receptor (which is recycled to the cell surface) and is delivered to a lysosome, a sac filled with digestive enzymes. Some of the enzymes break down the LDL's coat, exposing the cholesteryl ester core. Another enzyme clips off the fatty acid tails of the cholesteryl esters, liberating unesterified cholesterol, which leaves the lysosome. As we have indicated, all cells incorporate the cholesterol into newly synthesized surface membranes. In certain specialized cells the cholesterol extracted from LDL has other roles. In the adrenal gland and in the ovary it is converted into respectively the steroid hormones cortisol and estradiol; in the liver it is transformed to make bile acids, which have a digestive function in the intestine.

The amount of cholesterol liberated from LDL controls the cell's cholesterol metabolism. An accumulation of cholesterol modulates three processes. First, it reduces the cell's ability to make its own cholesterol by turning off the synthesis of an enzyme, HMG CoA reductase, that catalyzes a step in cholesterol's biosynthetic pathway. Suppression of the enzyme leaves the cell dependent on external cholesterol derived from the receptor-mediated uptake of LDL. Second, the incoming LDL-derived cholesterol promotes the storage of cholesterol in the cell by activating an enzyme called *ACAT*. The enzyme re-attaches a fatty acid to excess cholesterol molecules, making cholesteryl esters that are deposited in storage droplets.

Third, and most significant, the accumulation of cholesterol within the cell drives a feedback mechanism that makes the cell stop synthesizing new LDL receptors. Cells thereby adjust their complement of receptors so that enough cholesterol is brought in to meet their varying demands but not enough to overload them. For example, fibroblasts that are actively dividing, so that new membrane material is needed, maintain a maximum complement of LDL receptors (some 40,000 per cell). In cells that are not growing the incoming cholesterol begins to accumulate, the feedback system reduces receptor manufacture and the complement of receptors is reduced as much as tenfold.

Our observations in tissue cultures were confirmed when the receptor system was shown to have an important role in the body. Soon after we found the LDL receptor on cultured fibroblasts it was shown to be present on circulating human blood cells and on cell membranes from many different tissues of mice, rats, dogs, pigs, cows and human beings. The relative number of receptors and their functioning can be assessed in living animals and in human volunteers by injecting into the bloodstream LDL labeled with a radioactive isotope and measuring its rate of removal from the circulation. The rate has been shown to depend on the total number of LDL receptors displayed on all cells in the body. This can be demonstrated by modifying the apoprotein B-100 before the LDL is injected, so that it can no longer bind to receptors. James Shepherd and Christopher J. Packard of the University of Glasgow showed that the modified LDL circulates much longer than normal LDL.

Where is the LDL taken up? Daniel Steinberg of the University of California School of Medicine at San Diego and John M. Dietschy of the Health Science Center at Dallas have shown that in rats, rabbits, guinea pigs and squirrel monkeys about 75 percent of the receptor-mediated removal of LDL takes place in the liver. We have measured the number of receptors directly, in cell membranes isolated from different tissues. Most tissues are found to have some receptors, but those of the liver, adrenal gland and ovary—the organs with particularly large requirements for cholesterol—have the highest concentration of receptors.

What is the origin of circulating LDL? The mechanism of its production is more complex, and as yet less well understood, than the mechanism of its uptake and degradation. LDL is one component of the system that transports two fatty substances, cholesterol and various triglycerides, through the bloodstream. The fat-transport system can be divided into two pathways: an exogenous one for cholesterol and triglyceride absorbed from the intestine and an endogenous one for cholesterol and triglyceride entering the bloodstream from the liver and other nonintestinal tissues.

The exogenous pathway has been mapped by Richard J. Havel of the University of California School of Medicine at San Francisco and by others. It begins in the intestine, where dietary fats are packaged into lipoprotein particles called chylomicrons, which enter the bloodstream and deliver their triglyceride to adipose tissue (for storage) and to muscle (for oxidation to supply energy). The remnant of the chylomicron, containing cholesteryl esters, is removed from the circulation by a specific receptor found only on liver cells. This chylomicron-remnant receptor does not bind LDL or take part in its removal from the circulation.

LDL is a component of the endogenous pathway, which begins when the liver secretes into the bloodstream a large very-low-density lipoprotein particle (VLDL). Its core consists mostly of triglyceride synthesized in the liver, with a smaller amount of cholesteryl esters; it displays on its surface two predominant proteins, apoproteins B-100 and E, both of which can be bound by LDL receptors. When a VLDL particle reaches the capillaries of adipose tissue or of muscle, its triglyceride is extracted. The result is a new kind of particle, decreased in size and enriched in cholesteryl esters but retaining its two apoproteins; it is called intermediate-density lipoprotein, or IDL.

In human beings about half of the IDL particles are removed from the circulation quickly—within from two to six hours of their formation—because they bind very tightly to liver cells, which extract their cholesterol to make new VLDL and bile acids. Robert W. Mahley and Thomas L. Innerarity of the University of California School of Medicine at San Francisco have shown that the tight binding is attributable to apoprotein E, whose affinity for LDL receptors on liver cells is greater than that of apoprotein B-100. IDL particles not taken up by the liver remain in the circulation much longer. In time the apoprotein E is dissociated from them, leaving the particles, now converted into low-density lipoprotein (LDL), with apoprotein B-100 as their sole protein. Because of B-100's lower affinity for LDL receptors, the LDL particles have a much longer life span than IDL particles: they circulate for an average of two and a half days before binding to LDL receptors in the liver and in other tissues.

The central role of the LDL receptor in atherosclerosis was first appreciated when we showed that its absence is responsible for the severe disease called familial hypercholesterolemia (FH). In 1939 Carl Müller of the Oslo Community Hospital in Norway identified the disease as an inborn error of metabolism causing high blood cholesterol levels and heart attacks

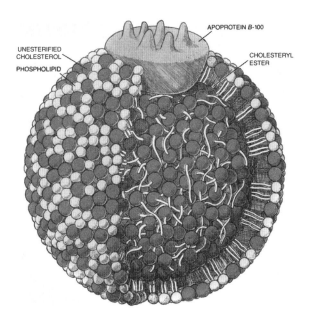

LDL, MAJOR CHOLESTEROL CARRIER in the bloodstream, is a spherical particle with a mass of three million daltons and a diameter of 22 nanometers (millionths of a millimeter). Its core consists of some 1,500 cholesteryl esters, each a cholesterol molecnle attached by an ester linkage to a long fatty acid chain. The oily core is shielded from the aqueous plasma by a detergent coat composed of 800 molecules of phospholipid, 500 molecules of unesterified cholesterol and one large protein molecule, apoprotein *B*-100. When blood cholesterol is elevated, increasing the risk of atherosclerosis, the reason is almost always an increase in circulating LDL.

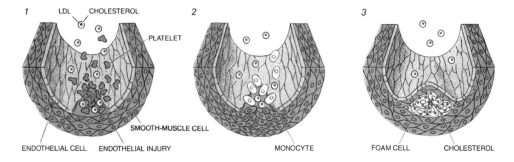

ATHEROSCLEROTIC PLAQUE develops slowly. Damage to the thin layer of endothelial cells that lines an artery initiates plaque formation. According to a model originally proposed by Russell Ross and John A. Glomset of the University of Washington School of Medicine, the damaged endothelium becomes leaky and is penetrated by low-density lipoprotein (LDL) particles and blood platelets (1). In response to the release of such hormones as platelet-derived growth factor, smooth-muscle cells in the layer below the endothelium multiply and migrate into the damaged area (2); at the same time white blood cells called monocytes invade the area and are activated to become scavenger cells called macrophages. The smooth-muscle cells and macrophages ingest and degrade LDL and become foam cells. If the blood LDL level is too elevated, cholesterol derived from the LDL accumulates in and among the foam cells. The accumulated cholesterol, cells and debris constitute an atheroma (3), which in time can narrow the channel of the artery and so lead to thrombosis.

in young people; he recognized that it is transmitted as a dominant trait determined by a single gene. In the 1960's Avedis K. Khacha-durian at the American University in Beirut and Donald S. Fredrickson at the U.S. National Heart and Lung Institute showed there are two forms of the disease, a heterozygous form and a more severe homozygous form. Heterozygotes, who inherit one mutant gene, are quite common: about one in 500 people in most ethnic groups. Their plasma LDL level is twice the normal level (even before birth) and they begin to have heart attacks by the time

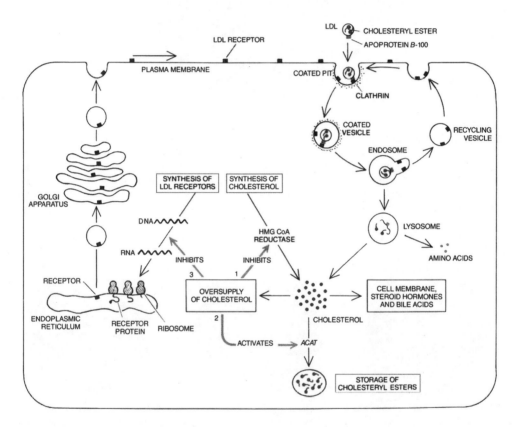

CIRCULATING LDL (*top right*) is taken into a cell by receptor-mediated endocytosis. LDL is bound by a receptor in a coated pit, which invaginates and pinches off to form a coated vesicle. Fusion of several vesicles gives rise to an endosome, in whose acidic environment the LDL dissociates from the receptor, which is recycled to the cell surface. The LDL is delivered to a lysosome, where enzymes break down the apoprotein B-100 into amino acids and cleave the ester bond to yield unesterified cholesterol for membrane synthesis and other cellular needs. The cellular level of cholesterol is self-regulating. An oversupply of cholesterol has three metabolic effects. It inhibits the enzyme HMG CoA reductase, which controls the rate of cholesterol synthesis (1); it activates the enzyme *ACAT*, which esterifies cholesterol for storage (2); and it inhibits the manufacture of new LDL receptors by suppressing transcription of the receptor gene into messenger RNA (3), which would ordinarily be translated on ribosomes of the endoplasmic reticulum to make the receptor protein.

they are 35; among people under 60 who have heart attacks, one in 20 has heterozygous FH.

If two FH heterozygotes marry (one in 250,000 marriages), each child has one chance in four of inheriting two copies of the mutant gene, one from each parent. Such FH homozygotes (about one in a million people) have a circulating LDL level more than six times higher than normal; heart attacks can occur at the age of two and are almost inevitable by the age of 20. It is notable that these children have none of the risk factors for atherosclerosis other than an elevated LDL level. They have normal blood pressure, do not smoke and do not have a high blood glucose level. Homozygous FH is a vivid experiment of nature. It demonstrates unequivocally the causal relation between an elevated circulating LDL level and atherosclerosis.

By what mechanism is the LDL level elevated? What is the particular function of the mutant gene? When we looked at cultured skin fibroblasts and circulating blood cells from FH homozygotes, we saw that the cells have either no functional LDL receptors at all or very few and therefore cannot bind, internalize and degrade LDL efficiently. The defective gene, in

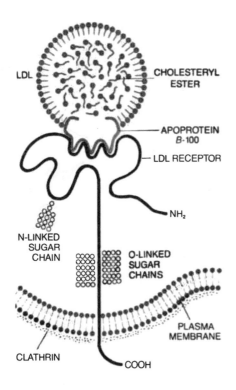

LDL RECEPTOR, a glycoprotein embedded in the plasma membrane of most body cells, was purified from the adrenal gland by Wolfgang J. Schneider in the authors' laboratory. David W. Russell and Tokuo Yamamoto cloned complementary DNA derived from its messenger RNA. The DNA's nucleotide sequence was determined and from it the 839-amino-acid sequence of the receptor's protein backbone was deduced. Sites of attachment of sugar chains to nitrogen (*N*) and oxygen (*O*) atoms were identified, as was a stretch likely to traverse the membrane. The actual shape of the receptor is not yet known; the drawing is a highly schematic representation.

other words, encodes the protein of the LDL receptor. Homozygotes, having inherited two defective receptor genes, cannot synthesize any normal receptors. The cells of FH heterozygotes have one normal receptor gene and one mutant gene; they synthesize half the normal number of receptors and can therefore bind, internalize and degrade LDL at half the normal rate.

Although all FH patients studied to date have a mutation in the gene encoding the LDL receptor, the mutations are not always the same. Depending on the particular site that has undergone mutation, the receptor may not be synthesized at all or it may be synthesized but then fail to be transported to the cell surface, fail to bind LDL or fail to cluster in coated pits.

Studies with radioactively labeled LDL show that the particles survive in the bloodstream of FH homozygotes about two and a half times as long as they do in people with a normal LDL-receptor gene. (Eventually the LDL is removed from the circulation by alternate but much less efficient pathways.) The predictable slowdown in the removal and breakdown of LDL is one major reason for the extremely high LDL level characteristic of FH, but it does not account for the entire rise. In addition to degrading LDL more slowly, a person homozygous for FH actually produces about twice as much LDL per day as a normal person. How can a defect in the LDL receptor lead to the overproduction of LDL? The answer to this question came from studies of a remarkable strain of rabbits with a genetic defect resembling the one in human FH.

The rabbits were discovered in 1978 by Yoshio Watanabe of the Kobe University School of Medicine and are called WHHL rabbits (for "Watanabe heritable hyperlipidemic"). They are homozygous for a mutant LDL-receptor gene and produce less than 5 percent of the normal number of receptors; they have high circulating LDL from the time of birth and develop atherosclerosis leading to heart attacks by the age of two. Studies done by us in collaboration with Toru Kita and David W. Bilheimer and by Steinberg and his colleagues showed that the rabbits, like their human counterparts with homozygous FH, make too much LDL as well as taking too long to break it down.

To learn the reason for LDL overproduction, Kita injected radioactively labeled VLDL, a precursor of LDL, into WHHL rabbits and normal animals and tracked the radioactivity through the fat-transport pathway. He found that triglyceride was removed from the VLDL, generating IDL, at the same rate in both groups. In normal rabbits the vast majority of the IDL particles disappeared rapidly from the circulation as they bound to LDL receptors on liver cells. In the WHHL rabbits, however, the liver cells lack LDL receptors, and so more IDL particles remained in the circulation and were eventually converted into more than the normal amount of LDL. In other words, a reduction in receptors has two effects in the rabbits—increased production and decreased removal of LDL—that act synergistically to raise the LDL level, which therefore rises disproportionately. Nicholas B. Myant and his colleagues at Hammersmith Hospital in London have shown the same thing is true in FH homozygotes.

Knowledge of the receptor deficiency in FH suggested a way to help the large number of patients with the heterozygous form of the disease. Perhaps we could stimulate the heterozygote's one normal gene to direct the synthesis of twice as many receptors as usual and so provide the patient with a normal complement of functional receptors. The possibility of such treatment was raised by something we had learned from cultured skin fibroblasts, namely that the feedback regulation of receptor synthesis takes place at the level of transcription. An excess of cholesterol reduces transcription of the LDL-receptor gene into messenger RNA, the nucleic acid that is subsequently translated by the cell's protein-synthesizing machinery to make the receptor protein; a cholesterol deficiency stimulates transcription and thus steps up the manufacture of receptors. We found we could get cultured cells from FH heterozygotes to make a normal number of LDL receptors (by making more messenger-RNA molecules from their single receptor gene) when we reduced the amount of cholesterol in the culture medium. How might we create an analogous cholesterol deficiency in the FH patient?

The liver takes up and degrades more cholesterol than any other organ because of its large size and its high concentration of LDL receptors. The bile acids into which most of the cholesterol is converted are secreted into the upper intestine, where they emulsify dietary fats. Having done their work, the bile acids are not simply excreted, however; they are largely reabsorbed from the intestine, returned to the bloodstream, taken up by the liver and again secreted into the upper intestine. This recycling of bile acids ordinarily limits the liver's need for cholesterol. We reasoned that if the recycling could be interrupted, the liver would be called on to convert more cholesterol into bile acids and this should lead the liver cells to make more LDL receptors.

A class of drugs that interrupt the recycling of bile acids was already known. They are the bile-acid-binding resins, gritty polymers carrying many positively charged chemical groups. Taken orally, these resins bind to the negatively charged bile acids in the intestine; because the resins cannot be absorbed from the intestine, they are excreted, carrying the bound bile acids with them. The first bile-acid-binding resin, cholestyramine, was synthesized more than 20 years ago and was found to lower the blood LDL level by an average of 10 percent. (A recent 10-year prospective study done by the National Heart, Lung, and Blood Institute indicated that such a reduction was enough to cut the incidence of heart attacks in a test group of middle-aged men by 20 percent.) What we had learned about LDL metabolism provided the missing rationale for such results: the interruption of bile-acid recycling increases the number of LDL receptors on liver cells.

The 10 percent drop in LDL level attainable with cholestyramine and other such resins was encouraging, but clearly a more profound reduction is necessary for treating FH heterozygotes. The limited efficacy of the resins stems from the dual response of the liver to a cholesterol deficiency. In addition to making more LDL receptors the liver increases its manufacture of HMG CoA reductase and makes more of its own cholesterol. We reasoned that this increased de novo synthesis of cholesterol partially satisfies the resin-induced demand for more cholesterol and so prevents the liver from maximally increasing the number of LDL receptors.

We thought inhibition of cholesterol synthesis might force the liver to rely more on LDL uptake and thus stimulate greater production of receptors. To block cholesterol synthesis we took advantage of the

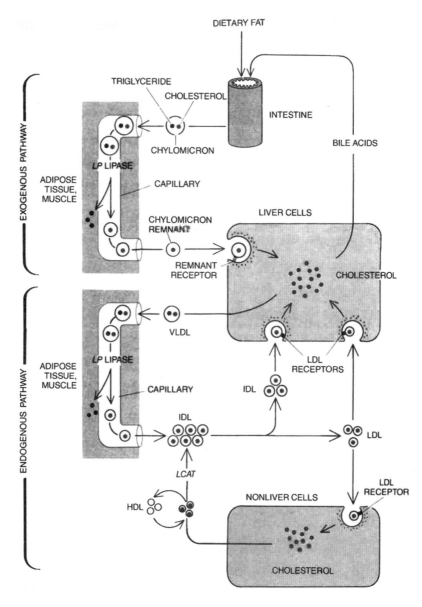

EXOGENOUS AND ENDOGENOUS fat-transport pathways are diagrammed. Dietary cholesterol is absorbed through the wall of the intestine and is packaged, along with triglyceride (glycerol ester-linked to three fatty acid chains), in chylomicrons. In the capillaries of fat and muscle tissue the triglyceride's ester bond is cleaved by the enzyme lipoprotein (LP) lipase and the fatty acids are removed. When the cholesterol-rich remnants reach the liver, they bind to specialized receptors and are taken into liver cells. Their cholesterol either is secreted into the intestine (mostly as bile acids) or is packaged with triglyceride in very-low-density lipoprotein (VLDL) particles and secreted into the circnlation, inaugurating the endogenons pathway. Again the triglyceride is removed in fat or muscle, leaving cholesterol-rich intermediate-density lipoprotein (IDL). Some IDL binds to liver LDL receptors and is rapidly taken np by liver cells; the remainder stays in the circulation and is converted into LDL. Most of the LDL binds to LDL receptors on liver or other cells and is removed from the circulation. Cholesterol leaching from cells binds to high-density lipoprotein (HDL) and is esterified by the enzyme LCAT. The esters are transferred to IDL and then LDL and are eventually taken up again by cells.

discovery by Akira Endo, now of the Tokyo University of Agriculture and Technology, of a remarkable natural inhibitor of HMG CoA reductase. In 1976 he isolated from a penicillin mold a substance called compactin. A side chain of the compactin molecule closely mimics the structure of the natural substrate of HMG CoA reductase, and so it binds to the enzyme's active site and inhibits the enzyme's activity. Alfred W. Alberts of the Merck Sharp & Dohme Research Laboratories and his colleagues isolated from a different mold a structural relative of compactin, called mevinolin, that is an even more potent enzyme blocker. Compactin and mevinolin were shown, by Endo and Alberts respectively, to lower the blood LDL level in animals. If our idea was correct, the drugs should be even more effective in conjunction with a bile-acid-binding resin.

In collaboration with Petri T. Kovanen we administered a bile-acid-binding resin to dogs either alone or along with one of the enzyme inhibitors. After two weeks we assessed the number of LDL receptors by measuring the ability of biopsied liver membranes to bind radioactive LDL. We found, as expected, that the resin alone generated a modest rise in the number of receptors. When the enzyme inhibitor was given too, the number of receptors rose much more. At the whole-body level this led to a marked increase in the rate of removal of LDL from the circulation. Together the two drugs caused a remarkable 75 percent decline in the dogs' LDL level.

With Bilheimer and Scott M. Grundy we went on to administer a resin and mevinolin to patients with heterozygous FH. Their LDL level fell by approximately 50 percent, into the normal range. Tests with radioactive LDL showed the drop was caused by an increase in LDL receptors. The single normal gene had been made to work twice as hard as usual, producing enough receptors to allow LDL to be removed from the circulation at a normal rate.

As might be expected, FH homozygotes, lacking even one normal receptor gene, do not respond to this two-drug treatment. Another approach must be found if they are to be helped. Thomas E. Starzl of the University of Pittsburgh School of Medicine has tested a surgical approach, following up on a suggestion that the homozygote's lack of receptors might be partially corrected if the patient could be given a liver from a normal donor.

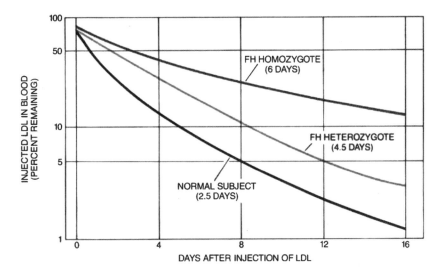

NUMBER OF LDL RECEPTORS in the body is assessed by injecting LDL labeled with a radioactive isotope and measuring the amount of radioactivity in blood samples for several weeks; the loss of radioactivity reflects the cellular uptake of LDL and hence the number of LDL receptors. The curves trace the removal of LDL from the circulation in patients with the homozygous and heterozygous forms of familial hypercholesterolemia (FH) and in normal subjects. In each case the mean life span of an LDL particle is shown in parentheses.

He transplanted the liver of a child killed in an accident into a six-year-old girl suffering from severe homozygous FH. (The patient had already had several heart attacks and her heart was so weakened that a heart transplant was necessary at the same time.) More than six months after the operation the patient was maintaining a total blood cholesterol level in the range of 300 milligrams per deciliter, compared with a preoperation level of about 1,200. Obviously liver transplantation is not an ideal treatment, but the results to date make it clear that receptors on the cells of the transplanted liver are functioning to remove LDL from the circulation.

What about the vast number of people in Western industrial societies who suffer heart attacks or strokes without having any genetic defect in the LDL receptor? Is what we have learned about FH relevant to the high incidence of atherosclerosis in the general population? We believe it is. The LDL-receptor hypothesis states that much of the atherosclerosis in the general population is caused by a dangerously high blood level of LDL resulting from failure to produce enough LDL receptors. The inadequate number of receptors can be attributed to subtle genetic and environmental factors that limit receptor manufacture even in people without FH. One environmental factor is a high dietary intake of cholesterol and of saturated fats derived from animal tissues.

Epidemiologic surveys done in many countries over the past 30 years have uniformly shown that atherosclerosis becomes severer as the mean LDL level rises in a population. As long ago as 1958 Ancel Keys of the University of Minnesota Medical School studied populations, in seven countries, in which the mean total cholesterol level varied from a high of 265 milligrams per deciliter to a low of 160. (He did not measure LDL cholesterol specifically, but because the level of lipoproteins other than LDL does not vary much, one can assume that the variations in total cholesterol reflected differences in LDL level.) Keys recorded the cholesterol level of 12,763 age-matched men in the seven countries, and 10 years later he determined which of the men had had a heart attack.

Two variables were found to correlate strongly with cholesterol level: the incidence of coronary atherosclerosis (as measured by fatal heart attacks) and the dietary intake of animal fats. In two villages (in Japan and Yugoslavia) where the mean total cholesterol level was 160 the incidence of fatal heart attacks was less than five per 1,000 men per 10 years. In eastern Finland, where the mean total cholesterol level was 265, the incidence of fatal heart attacks was 14 times as high. In populations with intermediate cholesterol levels (as in the U.S.) the incidence fell between the two extremes.

The correlation between cholesterol level and dietary intake of animal fats was even stronger than the correlation between cholesterol and atherosclerosis. Populations consuming small amounts of animal fats (as in Japan and Yugoslavia) had low cholesterol levels. Populations with a high intake of such fats (as in eastern Finland) had high levels. Subsequent studies of many different populations have confirmed Keys's findings: high LDL levels are the rule in populations that consume a large part of their calories as fats from meat and dairy products.

The LDL-receptor hypothesis provides a likely explanation of the epidemiologic data. A high average intake of cholesterol makes cholesterol accumulate in liver cells. The accumulation seems to be accentuated by ingestion of animal fats rich in saturated fatty acids. Even a modest accumulation of cholesterol in the liver would partially suppress the manufacture of LDL receptors. This could lead to an increase in the average LDL level that would be detectable in an entire population.

A nimal experiments by our group and by Mahley and Innerarity support the hypothesis that a high-fat diet reduces LDL receptors in the liver. In baboons, rabbits and dogs maintained on low-fat diets the number of LDL receptors is high and the animals degrade injected LDL rapidly; their LDL level is much lower than it is in human beings. When rabbits and dogs are fed diets high in cholesterol, their manufacture of receptors in the liver is suppressed by as much as 90 percent, and the result is a buildup of both IDL and LDL in the bloodstream. At birth human infants have LDL concentrations similar to those of other animal species; apparently newborn human beings make a large number of LDL receptors. During the childhood and early-adult years in industrialized societies, however, the LDL level rises three- or fourfold. Studies in adults injected with LDL suggest that the increase is attributable to a decrease in the number of receptors with age.

The causes of the acquired receptor deficiency in human beings are not all known. The high dietary intake of animal fats seems to be an important factor, but it is not the only one: even in people raised on diets extremely low in fats the LDL level tends to be higher than it is in other species. Such hormones as estradiol and thyroid hormone are known to stimulate the manufacture of LDL receptors in the liver, and it is possible that subtle abnormalities in these and other hormones contribute to the age-related decrease in receptors.

The concentration of LDL eventually attained in most middle-aged adults in the U.S. and in similar societies is associated by epidemiological data with accelerated atherosclerosis. Experiments with cultured cells show why. The receptors bind LDL optimally when it is present in the blood at a concentration below 50 milligrams per deciliter. The receptors in animals and in humans (judging by the LDL level in human infants) have apparently been selected by evolution to function at just such levels. Yet in Western industrial countries the average "normal" LDL level in adults is about 125 milligrams per deciliter, considerably above the concentration at which receptors bind LDL most efficiently.

One finding that is consistent with the LDL-receptor hypothesis has been reported by William R. Hazzard of the Johns Hopkins Hospital and his colleagues. They showed that ingestion by adults of a high-cholesterol diet (including three egg yolks per day) does lead to a decrease in the number of LDL receptors, which they measured directly in circulating lymphocytes. A definitive test of the hypothesis will, however, require a comprehensive and well-controlled study of the rate of metabolism of injected VLDL and LDL in members of populations with low-fat and high-fat diets and with varying LDL levels. That has not yet been done systematically.

If the LDL-receptor hypothesis is correct, the human receptor system is designed to function in the presence of an exceedingly low LDL level. The kind of diet necessary to maintain such a level would be markedly different from the customary diet in Western industrial countries (and much more stringent than moderate low-cholesterol diets of the kind recommended by the American Heart Association). It would call for total elimination of dairy products as well as eggs, and for a severely limited intake of meat and other sources of saturated fats.

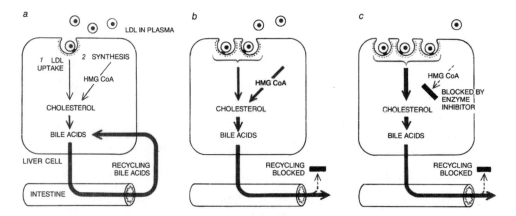

LIVER GETS CHOLESTEROL for conversion into bile acids from IDL and LDL taken up from the circulation (1) or by synthesizing it de novo (2). A key step in the long synthetic pathway is reduction of HMG CoA to mevalonic acid, a reaction catalyzed by the enzyme HMG CoA reductase. The enzyme is inhibited by the drugs compactin or mevinolin, whose side chain is so similar to that of HMG CoA that it blocks the enzyme's active site. Enzyme inhibition leaves liver dependent on uptake of IDL and LDL.

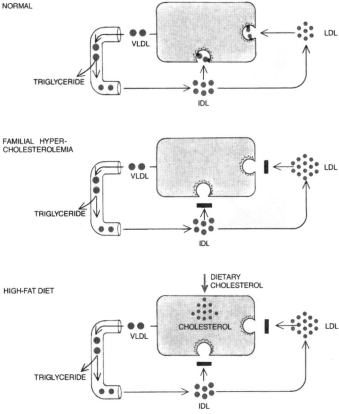

HETEROZYGOUS FH can be treated with a combination of drugs that stimulates manufacture of LDL receptors. Ordinarily the liver's demand for cholesterol is modified by the recirculation of bile acids (*a*). If the recirculation is prevented by a bile-acid-binding resin (*h*), more cholesterol is needed. Liver cells respond by increasing the number of LDL receptors, but also by increasing the rate of cholesterol synthesis. If a second drug is given to block enhanced synthesis (*c*), still more receptors are made and the blood LDL level is lowered.

LDL-RECEPTOR DEFICIENCY, whether genetic or acquired, has two synergistic effects that combine to raise the blood LDL level. VLDL secreted by the liver is converted into IDL in fat and muscle. In normal people about half of the IDL particles are taken up by LDL receptors on liver cells; the rest are converted into LDL (*top*). In FH (*middle*) a genetic defect diminishes the number of receptors on liver cells; an analogous deficiency is caused by diets that fill liver cells with cholesterol and so reduce receptor synthesis (*bottom*). In either case there are the same two consequences. IDL not taken up by liver cells remains in the circulation and is converted to yield increased amounts of LDL; the LDL in turn is removed more slowly.

We believe such an extreme dietary change is not warranted for the entire population. There are several reasons. First, such a radical change in diet would have severe economic and social consequences. Second, it might well expose the population to other diseases now prevented by a moderate intake of fats. Third, experience shows most Americans will not adhere voluntarily to an extreme low-fat diet. Fourth, and most compelling, people vary genetically. Among those who consume the current high-fat diet of Western industrial societies, only 50 percent will die of atherosclerosis; the other 50 percent are resistant to the disease.

Some individuals resist atherosclerosis because their LDL level does not rise dangerously even though they consume a high-fat diet; they may inherit genes that somehow circumvent the usual feedback system and maintain receptor manufacture at an adequate level. Barbara V. Howard of the National Institutes of Health Clinical Research Center in Phoenix has shown, for example, that Indians of the Pima tribe have relatively large numbers of LDL receptors, and maintain low LDL levels, in spite of a high-fat diet. In other individuals the arteries apparently resist the damaging effects of elevated LDL. For example, 20 percent of men with heterozygous FH do not have a heart attack before the age of 60 even though their blood LDL is very high.

G iven these reasons for constraint, what can be done to prevent accelerated atherosclerosis? One approach is to individualize dietary recommendations. A diet moderately low in animal fats would seem to be prudent for most people. The diet proposed by the American Heart Association, for example, would reduce blood cholesterol levels by as much as 15 percent and should somewhat lessen the incidence of heart attacks. On the other hand, people who have a strong family history of heart attacks or strokes, and who may therefore be particularly susceptible to the damaging effects of LDL, might well be encouraged to follow a diet extremely low in cholesterol and saturated fats—even if their LDL level is near the mean "normal" level. One can hope additional research will identify factors that either sensitize people to the ill effects of LDL or protect them from those effects.

Finally, therapy with drugs that increase the number of LDL receptors may turn out to be appropriate for at least some people who do not have FH but in whom the number of receptors is reduced by diet or other factors. If it is shown that these drugs do prevent diet-induced suppression of receptors and if the drugs can be shown to be safe for long-term use, it may one day be possible for many people to have their steak and live to enjoy it too.

## ABOUT THE AUTHOR

**Michael S. Brown** (born 1941) was awarded the Nobel Prize in Physiology or Medicine in 1985 along with Joseph L. Goldstein for their discoveries concerning the regulation of cholesterol metabolism. Dr. Brown holds the W. A. (Monty) Moncrief Distinguished Chair in Cholesterol and Arteriosclerosis Research, the Paul J. Thomas Chair in Medicine, and is a Regental Professor of the University of Texas. He also is a member of the Board of Scientific Directors at The Scripps Research Institute.